The Science o

The Science of Sci-Fi Cinema

Essays on the Art and Principles of Ten Films

Edited by
VINCENT PITURRO

McFarland & Company, Inc., Publishers
Jefferson, North Carolina

LIBRARY OF CONGRESS CATALOGUING-IN-PUBLICATION DATA

Names: Piturro, Vincent, 1969– editor.
Title: The science of sci-fi cinema : essays on the art
and principles of ten films / edited by Vincent Piturro.
Description: Jefferson : McFarland & Company, Inc., Publishers, 2021 |
Includes bibliographical references and index.
Identifiers: LCCN 2021024157 | ISBN 9781476683300 (paperback : acid free paper) ∞
ISBN 9781476641232 (ebook)
Subjects: LCSH: Science fiction films—History and criticism. |
Science in motion pictures. | BISAC: PERFORMING ARTS /
Film / Genres / Science Fiction & Fantasy | SCIENCE /
Essays | LCGFT: Science fiction films.
Classification: LCC PN1995.9.S26 S635 2021 | DDC 791.43/615—dc23
LC record available at https://lccn.loc.gov/2021024157

BRITISH LIBRARY CATALOGUING DATA ARE AVAILABLE

**ISBN (print) 978-1-4766-8330-0
ISBN (ebook) 978-1-4766-4123-2**

© 2021 Vincent Piturro. All rights reserved

*No part of this book may be reproduced or transmitted in any form
or by any means, electronic or mechanical, including photocopying
or recording, or by any information storage and retrieval system,
without permission in writing from the publisher.*

Front cover image © 2020 Shutterstock

Printed in the United States of America

*McFarland & Company, Inc., Publishers
Box 611, Jefferson, North Carolina 28640
www.mcfarlandpub.com*

To the memory of Brit Withey
(1968–2019)

Table of Contents

Introduction
    VINCENT PITURRO — 1

Chapter 1: *Arrival* — 11
    The Science of Communicating with Aliens
    KA CHUN YU — 16

    Arrival from a Linguist's Perspective
    ANDREW J. PANTOS — 23

Chapter 2: *Interstellar* — 31
    How Easy Is *Interstellar* Colonization?
    KA CHUN YU — 39

Chapter 3: *2001: A Space Odyssey* — 49
    Catching HAL
    KA CHUN YU — 59

Chapter 4: *Children of Men* — 73
    Infertility and the Near-Future
    NICOLE L. GARNEAU — 80

Chapter 5: *Perfect Sense* — 84
    All They Need to Know: Connectedness
    NICOLE L. GARNEAU — 90

Chapter 6: *Upstream Color* — 97
    Parasites and Defying Explanation
    NICOLE L. GARNEAU — 103

Table of Contents

Chapter 7: *Contact* — 110
  The Sounds of *Contact*
    NAOMI PEQUETTE — 117

Chapter 8: *Jurassic Park* — 123
  Jurassic Park and the Dinosaur Renaissance
    JOSEPH SERTICH — 131
  Monstrous Life Finds a Way: *Jurassic Park* and Monstrosity
    CHARLES HOGE — 137

Chapter 9: *King Kong* — 145
  Gigantic Animals
    JEFFREY T. STEPHENSON — 152
  Falling Off the Phallus of Civilization: On Max Steiner's Soundtrack to *King Kong*
    ROGER K. GREEN — 159

Chapter 10: *The Martian* — 167
  The Martian: Fact, Fiction or Fantasy? An Interview with Steven Lee
    KA CHUN YU — 174

About the Contributors — 187
Bibliography — 189
Index — 190

Introduction

VINCENT PITURRO

Science fiction has always inspired awe, wonder, surprise, as well as fear and anxiety. It is one of the great genres in the history of cinema, from the very beginning of the nascent art form to the current day of miraculous, CGI-heavy visual treats. It has the ability to take us to faraway galaxies or to stay very close to home; it has the ability to curate great curiosity and great trepidation; it has the ability for potential and for limitation; it has the ability to comment on our current society no matter the setting; and it has always wonderfully wedded technical prowess with great storytelling. It is a wonderful marriage of art and science.

But how much actual science is really in those science fiction films most of us know and love? The answers are as varied as the films themselves: none, a lot, and some. Each film and each filmmaker has a different project and rhetorical objective—some focus solely on story and move heaven and Earth to get there no matter the science; some hew close to science and refuse to go outside those lines; still others tread the line between art and science, guiding the narrative toward a happy medium. How each film fulfills its objective is their own choice, and every film has its singular goal. The fun is in how the film gets to that goal, how it organizes itself, how it sees the contemporaneous society as well as the future, and finally how we, as viewers, receive all of that information.

This collection analyzes ten films, one per chapter. I write the first part of each chapter—analyzing the film in the detail—and a scientist writes the second part of the chapter discussing the science in the film. A few of the chapters have a third, bonus section. This book project is the progeny of an annual Science Fiction Film Series in Denver, Colorado, started in 2010 with my late friend and colleague Brit Withey (Artistic Director for the Denver Film Society). Our initial goal was simple: to share the confluence of art and science with the community. We were both fans of science fiction cinema, and we thought filmgoers and museumgoers alike would be interested in/enriched by such a unique endeavor. We brought the idea to the Denver Museum of Nature and Science, and they signed on. The format was to present a series of films and pair each film with a scientist in their area of specialty: an astrophysicist for space films; a geneticist for films on genetic engineering; a paleontologist for films on re-engineered dinosaurs; a zoologist for films with animals; and so on. For that first year, we were worried about finding enough films to fit the bill and to which we could match scientists. Ten years later, the series is still going strong and the possibilities seem endless. After many years in front of interesting and enthusiastic crowds, we finally decided to put our ideas down in print.

First, in great Aristotelian fashion, we will define our terms, both in terms of science fiction and in terms of film analysis. The first and most elemental questions we will address are definitional: "What is science fiction?" and "What does the genre entail and what does it not?" Of course, these are not easy questions, and the only easy answer is that there are many answers. Science fiction may be akin to what Forrest Gump famously mused about a certain state of being: science fiction is as science fiction does. And we can drill down from there. One definition I always find useful is that sci-fi takes what is in the laboratories today and extrapolates out to what is not only probable, but what also might be *possible*. Another definition I find useful is the difference between sci-fi and horror, as posited by Film Professor and Historian Dr. Bruce Kawin: "When faced with a door to the unknown, horror closes it and runs, whereas science fiction opens it and explores." In addition, one of the defining characteristics for me is that science fiction should always have a basis/foundation in Earth. There should be implications for our own planet, in whatever state the planet may be in the film's setting. So following that definition, *Star Trek* fits into science fiction for me where *Star Wars* does not (I find it more science fantasy). And finally, the best definition of science fiction may be the term itself: a mix of science and fiction. No matter how you see it, however, part of the fun of defining the term is the journey itself. That road to defining the term may be more interesting than the end point—if there is one. Still, there are several things we can glean from the history of the genre itself.

One specific, defining characteristic of science fiction certainly has been its indelible comment on the contemporaneous society. From the very beginning, science fiction cinema always had *something to say*. It many respects, it is the Western genre in reverse: whereas Westerns said something about the contemporaneous society while setting them in the past, science fiction does the same but the setting is now the future. Both genres have been bedrocks on which cinema was built, and they have stood the test of time. Westerns may have dominated Hollywood in its early days, but science fiction has become increasingly popular and enduring as Westerns have fallen out of favor. And the history and influence of science fiction, just like the Western, goes all the way back to the initial days of the medium itself.

Science fiction was a part of cinema from its earliest days: magician and early film genius George Méliès' *A Trip to the Moon* was released in 1902 and it would become a landmark film in many ways, including groundbreaking special effects that Méliès was inventing right there on the spot. Since much of early film took from literature (and the stage), Jules Verne and H.G. Wells were big influences on Méliès. There were also early silent adaptations of other classics, such as *20,000 Leagues Under the Sea, Frankenstein*, and *Dr. Jekyll and Mr. Hyde*. Science fiction film took from science fiction literature, and that partnership remains strong today; of the ten films in this volume, five are directly taken from either a story or novel.

The early masters of cinema were all influenced by science fiction, or science, as well. Even though they didn't work specifically in sci-fi, Charlie Chaplin and Buster Keaton would oftentimes portray men as machines (to make a comic point or a symbolic point) and commented on the ramifications of the machine age in the late teens, '20s, and '30s. These ideas were shared with the era's sci-fi. Chaplin was also very much influenced by the first great science fiction masterpiece, *Metropolis* (1927), incorporating some of the ideas, tropes, and symbolic visuals into his brilliant film *Modern Times* (1936). Both Chaplin and Keaton were also very interested in film technology as a whole and advanced the medium of film with their own technological innovations. Keaton, especially, was a master at using two (or more) different shots in the same frame, achieved by covering the lens in one area, filming that area, rewinding the film, covering the lens in the area already filmed, and then filming in the empty space. The result: two different shots in the same frame. Such innovation would shape all of film, but sci-fi in particular.

Science fiction in all forms would continue to capture the popular psyche well into the new century. The pulp magazine serials and comic books of the '20s and '30s were very popular among mass audiences, and they were very much "soft sci-fi": simple narratives, fantastic stories, and less concerned with science than with story, mostly of the "fantastic narrative" type. Those pulps and comic books easily translated into early cinema: the stories of amazing gadgets, wild machines, and heroic scientists (not to mention space travelers) showed up in the serial films of Buck Rogers and Flash Gordon. Both became very popular and made science fiction a very popular genre of early cinema. Along with the Western, science fiction became a bedrock of the "pitchas" and drew big audiences—propelling the new art form.

It is also instructive to look at sci-fi in terms of decades and how sci-fi would come to represent and reflect each decade. The sci-fi of the '20s and '30s was primarily concerned with dystopias, Utopias, the machine age, and fascist societies. The world experienced the rise of the super-cities in the post-industrial age and then moved into the machine age—mechanization, Ford-ization, and the rise of machines taking the place of people (and conversely, people turned into machines in such places as assembly lines). The world of the '20s also featured a great debate between capitalism and communism (remaining a staple of the genre's thematic focus beginning with *Metropolis* and continuing—in other forms—to the present) that would eventually lead to regime changes, dictatorships, and finally, World War II and beyond. The world was a dynamic place during this point in our history, and many felt that the future was uncertain. Sci-fi was the perfect venue for examining all of these issues.

Post–World War II sci-fi changed as much as the world around it did. As J.P. Telotte notes in *Science Fiction Film*, there were several factors involved: sci-fi literature became more widespread, partly as a result of the genre's move into mainstream literature—the popularity of writers Ray Bradbury (*The Martian Chronicles*) and Isaac Asimov (*I, Robot*) exploded, and sci-fi became more respectable and less of a niche market. The postwar population also became an increasingly literate audience, with the GI Bill sending eager veterans to school, a newly prosperous economy giving everyone more leisure time, and an increasingly chaotic world leading people to seek out diversions. Sci-fi also experienced an expanded breadth of subject matter and the depth of treatment of such subject matter, as evidenced in the writings of Bradbury, Asimov, and others.

Political and literary consciousness was infused into the sci-fi literature for the first time as well. The '50s and '60s sci-fi literature evolved to include subjects that were on the minds of everyone after such traumatic decades: explorations of identity, failures, limits, ends, and the possible finality of our world. The war and the resulting unprecedented devastation, along with the unimaginable death toll around the world, took its toll on all aspects of society and manifested itself in all aspects of art. Science fiction film also saw a new focus on monsters and aliens, and even more fantastic narratives as we started to look to the skies more and more. Many of these representations were a function of UFO sightings of the late '40s (Roswell in 1948), environmental effects of the war (*Godzilla* [1954] and the Hollywood B-movies), and the threat of nuclear satellites circling the Earth. The fear and anxiety of that society were reflected in the science fiction films of the time, including *The Day the Earth Stood Still* (1951) and *Invasion of the Body Snatchers* (1956). Such films were also the beginnings of "adult sci-fi," with increasingly complex allegorical content.

Tracking along with the thematic aspects was an increase in film technology and special effects. The movies were exploding in the '40s and early '50s and studios were flush. Movie attendance was high, the business had figured out the inclusion of sound, and production exploded. In the world of sci-fi, cinema began to take on space exploration, and eventually, the challenge of President Kennedy in 1962 to put a man on the moon not only challenged NASA and scientists in general but inspired Hollywood as well. Even though the industry would soon struggle in the late '50s and '60s, space became a new frontier, and not just on film but TV as well. *Star Trek* debuted in 1966 and that original series lasted for three years, becoming a staple of TV and movies ever since. International science fiction also blossomed at this time; making the great sci-fi film became something of a holy grail for European art directors: Jean-Luc Godard tried his hand with *Alphaville* in 1965, François Truffaut made *Fahrenheit 451* in 1966, and the great Italian director Federico Fellini was always interested in making a science fiction film. He never did—more on that later.

The '70s was also a particularly verdant period for sci-fi, and in particular, political and ideological sci-fi found its way into the mainstream: environmental issues (*Planet of the Apes, Soylent Green, Logan's Run*), feminist films (*The Stepford Wives*), developing AI (*Westworld, Futureworld*), and even crazy science fantasy/fantasy films such as *Barbarella* and *Zardoz* brought audiences back to the theaters as the movies made a comeback. Writers and directors had more freedom in this period as the Hays Code was abolished and the studio system was in shambles.

The latter part of the '70s and into the '80s would see the end of that independent era and the return of the big studios, including the advent of the summer blockbuster. Films such as *Jaws* (1975) and *Star Wars* (1977) would usher in this era, and big-budget sci-fi would return (especially on the heels of the *Star Wars* box-office success). This wonderful period in the history of cinema would come to an end, but new films and new technology would soon make sci-fi increasingly mainstream and even more popular.

The late '70s and '80s saw a turn from the overtly political films of the '60s and early '70s; this dynamic was another direct result of the studios re-taking control of the industry. The era would also see the rise of the sci-fi epics and mainstream films—*E.T.* (1982) and *Close Encounters of the Third Kind* (1977), for example. The other important development at this point was the continuing advancement in technology. The *Star Wars* films would move the bar for special effects, and then *Jurassic Park* in 1993 would mark a sea change in the industry with the advent of CGI and the waning days of the "puppets" and mechanical special effects of the '70s and '80s. Soon, CGI would dominate the industry and the sci-fi genre in particular.

On the thematic side, as the Cold War–era waned, the oppositional structure of capitalism vs. communism would morph into a singular indictment of hyper-capitalism and corporatism. We also see a move toward the postmodern during this period as well, with the impact of machine intelligence, AI, and robotic automation. *The Terminator* (1984), and *Robocop* (1987) would be iconic sci-fi of the '80s and push the special effects and thematic aspects along. In the '90s, we see the big jump in CGI with *Jurassic Park* and then films such as *Dark City* (1998) and *The Thirteenth Floor* (1999) dealing with the subject of alternative realities. All of these thematic and cinematic changes would lead up to *The Matrix* in 1999. That groundbreaking film—in both special effects and in its vision of the future—would take everything simmering in sci-fi up to that point and turn it up to a full boil. While *The Matrix* would certainly change the direction of the genre, the events of 9/11 caused a sea change. The predominant opposition of sci-fi for several decades had been humans vs. machines (ever since *2001: A Space Odyssey* in 1968), but the post–9/11 films would reconstruct the thematic focus: films would no longer be set in a faraway future that kept us at a safe distance from all of the issues therein (*Alien* or *The Matrix*). The films would become closer, "Near-Sci-fi" as a I call it, and the issues would be more urgent. The central problem also became simpler: *us*. We no longer needed an "Other" when the Other was us. The cinematic focus moved from CGI, video-game-like films to a more realistic aesthetic in this period as well. The new sci-fi would take off from there.

Aside from the historical review of sci-fi, the other Aristotelian task we have is to define our terms for film analysis. I will use these terms in each chapter to engage in cinematic analysis as well as thematic analysis. Four terms form the building blocks of our study:

1. *Mise-en-scène*—literally "placed in the shot," covers everything placed in front of the camera before shooting. This includes settings, subjects, and compositions. Things such as objects in the frame (a desk, chairs, an ashtray, an actor, a house, etc.); or settings such as inner city, country, desert, mountain, indoor, outdoor; or even the way the certain objects are framed: close, tight, far away, etc., hold meaning for the scene.

2. *Cinematography*—includes the camera, the lighting, and the medium on which the movie is made (film or digital). This can be a very technical aspect of film and it includes the color (black and white or color film stock); digital camerawork; the angles (low-angle, high-angle, point-of-view, bird's-eye-view); the distance from subject (extreme long shot, long shot, medium shot, close-up); and the intensity and angle of the lighting (high-key vs. low key, hard vs. soft, top lighting, main light, etc.). It can also include the overall style of shooting, such as realism vs. expressionism.

3. *Editing*—consists of the way the shots are put together. Continuity editing provides seamless transitions from shot-to-shot so the viewer understands where the subjects are and in what order the events are taking place. Other types of editing (such as dialectical montage or jump cutting) may be reflexive, disjunctive, or suggestive of something in particular. The most common form of editing is the "cut" which means one shot ends and the other begins. Where specific areas of editing are discussed, such as cross-cutting, I define them and place them in context.

4. *Sound*—includes dialogue, music, sound effects, and silence. Sound can be either diegetic (emanates from within the frame) or non-diegetic (which comes from outside the frame, such as music on the soundtrack or voice-over). Dialogue is any sound made by the human voice; sound effects are sounds *not* made by the human voice (a door slamming, a car screeching, a monster roaring, etc.). And silence is golden.

I will go into more detail about each of these aspects in the individual chapters and apply them to the films along the way. And where needed, I will illuminate specific terms/concepts/ideas that need extrapolation. Still, you do not need a degree in film studies to follow along. You will be given everything you need to know.

That brings us to the individual films and chapters. They were all chosen in discussion with each of the scientists, and they flow from our work in the annual Science Fiction Film Series. Some are more popular/well-known films, and some are smaller, independent films. One of the more fascinating aspects of the series over the years has been the eclectic mix of film types, and we have tried to capture that in these pages.

The first chapter covers one of the best science fiction films of the past decade, *Arrival* (2016). This is one of the chapters with three sections: the film analysis written by me; the science, written by Dr. Ka Chun Yu from the Denver Museum of Nature and Science; and the third section is written from a linguist's point of view—considering the main character is a linguist communicating with aliens, it is a fascinating perspective from Dr. Andrew J. Pantos of the Metropolitan State University of Denver.

The second chapter covers another of the best and most interesting science fiction films of the past decade, *Interstellar* (2014). It is a sprawling, dense, and wonderfully complex film on many levels. It is also chock-full of science, with its own staff of scientists who worked on it. Dr. Ka Chun Yu writes the science section and concentrates on the possibility of establishing off-world colonies, which is the overall goal of the narrative.

The third chapter analyzes perhaps the best and most famous science fiction film of all time—Stanley Kubrick's *2001: A Space Odyssey* (1968). I analyze the film but also situate it in the larger context of cinema history; considering that cinema was at a low point during the '60s, the film began the upward trajectory of the entire industry. Dr. Ka Chun Yu focuses on artificial intelligence in his section of the chapter.

Chapter 4 starts a section on smaller, independent films with *Children of Men* (2006), but it also enters the area where sci-fi changed in the wake of 9/11. I discuss the cinematic elements of the film, and in addition, how it highlights the change in both the thematic focus and the cinematic focus during this era. Geneticist Dr. Nicole L. Garneau handles the science and focuses on infertility and the possible reasons for it in the film as well in our world.

Chapter 5 examines a smaller film with big stars and big ideas, *Perfect Sense* (2011), and follows in the trend of post–9/11 films described in the previous chapter. Dr. Nicole L. Garneau discusses the science of senses and how that informs the narrative. The film is a great example of how science fiction, in any form, style, or type, always gets to the essential question I see at the center of the genre.

Chapter 6 covers a very independent film in *Upstream Color* (2013), one with a fascinating and perhaps, oblique, narrative that challenges and delights on many levels. I argue that it is a formalist gem—one in which all of the cinematic elements point toward wonderful thematic depth. It is a great example of how we have tried to include all types of films, from the very biggest to smaller, more obscure films that may not be so well-known but are fascinating and important nonetheless. Dr. Nicole L. Garneau speaks to the science of parasites, among other things, in dissecting the science in the film.

Chapter 7 begins a section on big-budget, Hollywood films with one of the best big-budget Hollywood directors, *Contact* (1998) from director Robert Zemeckis. Based on a book by Carl Sagan, it is a sprawling film with a huge cast and huge ideas. I try to reign in the narrative and also discuss the movement toward CGI at this point in cinema history. Space Scientist Naomi Pequette from the Denver Museum of Nature and Science examines the science in the film, including listening for sounds of life in the universe.

Chapter 8 covers one of the most influential films in the last half of the 20th century, *Jurassic Park* (1992), and how the Steven Spielberg film helped move special effects from the era of puppets to the era of CGI. It is a groundbreaking film but also deliciously entertaining. Paleontologist Dr. Joseph Sertich from the Denver Museum of Nature and Science examines the science of the film, and yes, answers the question everyone is dying to know: "Can we bring dinosaurs back?" Dr. Charles Hoge from

Metropolitan State University of Denver adds a third section to the chapter on how *Jurassic Park* fits into the history of science fiction monster movies.

Chapter 9 examines a giant in the history of science fiction monster movies and one of the most influential films of all time—in terms of special effects—in Merian C. Cooper's *King Kong* (1933). I detail all the fascinating cultural and historical aspects of the film as well as the groundbreaking special effects. Zoologist Dr. Jeffrey T. Stephenson from the Denver Museum of Nature and Science discusses the animals in the film. Dr. Roger K. Green from Metropolitan State University of Denver provides a detailed look into the music in the film, also a very famous aspect, in a third section of the chapter.

And finally, Chapter 10 is a treat: after my analysis of *The Martian* (2015), Dr. Ka Chun Yu gives an interview with Dr. Steven Lee, a prominent Mars scientist from the Denver Museum of Nature of Science, discussing the actual science in the film—or perhaps the lack thereof. The analysis is the very heart of what the entire book strives to achieve: the marriage of art and science, and our wish to bring that to a larger audience in an interesting and informative package.

If there is one controlling factor in my own analysis of each film, in each chapter, it is to look at both the cinematic and thematic aspects of each film. I also try to give a sense of the historical, cultural, and cinematic contexts of each film, and I include a background of each director and how they came to the project. And finally, I strive to address how each film answers the central question of science fiction, in my opinion: "What does it mean to be human?" I hope the ensuing pages may do exactly what science fiction always done: inspire awe, wonder, surprise, and even fear and anxiety. All very human emotions.

Chapter 1: *Arrival*

Amy Adams as Dr. Louise Banks reaches out to the aliens in *Arrival* (Paramount Pictures, 2016).

Arrival (2017) is based on the 1998 science fiction short story "Story of Your Life" by Ted Chiang—a winner of several prestigious sci-fi literary awards, including the Nebula Award for Best Novella. It tells the story of a linguist who helps the U.S. military communicate with extraterrestrial life forms that have landed on our planet, one of 12 ships that materialized on Earth at the same time in different locations around the world. The entire endeavor starts out as a worldwide, international effort to communicate with the aliens, but soon the military (the U.S. and others) intervenes. The story as laid out this way is very straightforward, but the film is hardly so. The essence of the film is in its structure, characterizations, and overall humanistic view of the world. In one sense, it is about aliens coming to Earth. In its more basic, important, and alluring sense, it is about *us*—who we are and what we are. It speaks to the most basic question of science fiction, in my opinion: "What does it mean to be human?" That central question of science fiction will be a constant throughout the book, and each film answers it in different, and wonderful ways. We arrive with *Arrival*.

Directed by Denis Villeneuve, a Canadian who also directed *Blade Runner 2049, Sicario, Prisoners,* and *Incendies,* it breaks down some of the facile tropes of blockbuster sci-fi and speaks to an informed and intelligent audience. No stranger to science fiction, Villeneuve states: "The main thing I was attracted to was this idea of exploring culture shock, exploring communication, exploring this idea of language changing the perception of your reality. That was gold." When Chiang was presented with the idea of an adaptation, he was initially skeptical. He didn't think the cerebral, interior-based emotional underpinnings of his work would relate to the screen. The producers enticed him with the presence of Villeneuve, a director known for very unique films with arresting visuals and sound that still created complex characterizations. They sent Chiang a copy of Villeneuve's *Incendies* (2010), the indie hit that tells the story of twins who wish to uncover their family history in the middle east. Chiang was impressed by the thoughtful and arresting film and also by the producers' choice to send him that particular film. He was on board.

Amy Adams (Louise), for her part, was attracted to the role of an intellectual female character who goes toe-to-toe with the males, but the project also interested her as a mother—since she has a young daughter of her own. Adams recalls that she was not interested in working at the time and just wished to take a break and be a mother. But after she read the script, she was hooked, and she decided to take the break *after* making the film. Jeremy Renner (Ian) took the part because he loved the script, but he also wanted to work with Amy Adams. He especially loved the woman-centered story and how his character was just secondary to that character. As Adams said, "Not many men would take the part of playing second to a lead played by a woman. Jeremy was fantastic." Her comment speaks not only to the communal and collaborative nature of the film or to the filmmaking process in general, but it is also a sad comment on the movie business itself—how it is still so patriarchal and male-ego-driven in spite of its intermittent and sometime progressive tendencies, especially in the light of the Me-Too movement. The story of a strong woman in the lead role is also quite rare in the history of sci-fi; you would have to go back to *Alien* to find an antecedent, without much else in-between. Considering the overall talent and experience of the entire production, the cast and crew were set up for success from the start.

The film was well-received critically and it did well at the box-office, which is not always the case for a heady sci-fi film. It was nominated for Eight Academy Awards, including Best Picture and Best Director, and it won a single Oscar, for Best Sound Editing. Sci-fi is not always recognized at the Academy Awards, and the inclusion of the film outside the technical categories is a testament to the wonderful production as a whole. It is not just a great sci-fi film, it is a great film, period. When looked at as a whole, the run of intellectual sci-fi in the 20-teens is quite verdant, considering the critical and box-office successes of this film and others such as *Interstellar* (2014) and *The Martian* (2015). All of these share certain characteristics with *Arrival*: they assume an intelligent audience, they are not afraid to tackle big questions, they combine in-depth thematic discussion with technical and cinematic prowess, and they at once feel the history of science fiction cinema while pointing forward toward a new future for the genre. Most of all, they address what I see as the essential question of science fiction and perhaps all of our lives: "What does it mean to be human?"

The film addresses this monumental question in many ways, and quite thoroughly through its thematic depth and cinematic aspects. There is a wonderful continuity and confluence among all of these aspects that brings the film together in wonderful ways. Beyond the inspirational source material, the excellent adaptation of story to script, the steady and inventive helm of Villeneuve, and the intelligence and depth of the performances, the interiority of the original story bubbles to the surface with sensitivity, intellect, and verve.

The marriage of thematic depth with the cinematic aspects comes through clearly in the symbolism, the most prominent of which is the circle, for many reasons. The circle is the deconstruction of the linear—it is a visual representation of the physical properties that Chiang had in mind when he wrote the short story. Chiang was thinking about physics, and properties that speak to a more non-linear way of thinking rather than a simple cause-and-effect mode of existence. What if we knew the beginning and end point of something, but then must fill in the rest? What if causes and effects didn't matter as much as the choices we make along the journey? What if the way we look at the world, and the way the physical word is organized, is more circular than linear? Chiang wished to explore these ideas in a speculative narrative, and they take form, initially, in the shape of a circle.

The circle has several thematic implications in the film: as the representation of nonlinear time, as the way in which the aliens communicate through their language, and as the way in which the actual narrative of the film is structured. The first instance we get of this structure is in the first scene of the film, which, when you go back and mentally re-construct the narrative, is actually the last scene of the film in a linear timeline. In other words, the first scene of the film takes place after the narrative told starting with the second scene. So in that second scene, the aliens arrive, and Louise is subsequently engaged by the military to communicate with them. This is where she meets Ian and they begin working together. They figure out the alien language, communicate with them, and as the other 11 crews around the world working on similar projects find, everything is soon hijacked by the military and certain destruction is imminent. Louise, however, is able to thwart the military intervention by speaking to the

Chinese military leader and talking him down. She does so by repeating to him something told to her (by him) in a future meeting (in other words, a meeting that takes place in the future). Louise was given the gift of nonlinear time/thinking/experiences by the aliens and can see not only beginning points of events (the present), but also the ending points of events (the future). The aliens had come to earth to essentially ask for help in thwarting their own demise many years in the future, and Louise is the conduit for that human education. Such is the structure of the narrative; it is nonlinear, it is asynchronous, and it is fraught with both awesome power and existential crises. As Louise finds out, she and Ian will have a daughter who will become sick and die young. Louise decides to still have the child (without telling Ian the eventuality of her death) and enjoy the time with her daughter. Here is the moral center of the film, as Chiang puts it: "If you knew your child was going to die, and there was nothing you could do about it, would you still have that child?" This choice, put in such stark simplicity, is one of those moments that raises the question "What does it mean to be human?" That choice, given the knowledge, speaks so eloquently to the beauty and the peril of our humanity.

Everything flows from this choice and the visual representation of it. The *mise-en-scène* is infused with circles that speak to the circularity of our existence and our choices. The inky representations of the alien language are formed in a circular pattern that is at first difficult to comprehend—when we base it on our own formation of linear language structures. But once Louise sees the patterns and is able to translate, she is given the gift (a "tool") of seeing/thinking in terms of beginnings and endings, rather than causal relationships. The circular language representations are each a story, told with knowledge of the journey and the plot points. One of the fascinating aspects of this type of communication is that arriving at the endpoint doesn't preclude the intervening events. In other words, there is no oversimplified time-travel element that allows Louise to "fix" her daughter and arrive at a different conclusion. Her daughter still dies, and it is not so such much the ending event that is most important, or most devastating in this case, but the events leading up to it. Louise enjoys, relishes, and *chooses* the journey.

This process is also akin to the filmmaking process. While a film is scripted, sometimes visualized (storyboards or otherwise), and engineered toward a specific end, the process itself is messy and unpredictable. There are a gaggle of people involved in the process from start to finish: screenwriters, producers, production designers, cinematographers, editors, sound designers, actors, directors, and countless other cast and crew. Nobody knows what will happen on that journey, or what the finished product will look like (even with a fixed ending/endpoint), but the filmmaker chooses to engage in that process and take that journey. Along the way, one can make beautiful discoveries and/or encounter unthinkable disasters. Sounds like life: we all know we are going to die, but how we live is our choice. Most of us choose to take that journey.

The *mise-en-scène* also speaks to the messiness of the journey. The juxtaposition of the spaces inside and outside the tent city around the alien ship are the perfect example. Inside the tents, the space is cluttered, confining, and ultimately claustrophobic. The compositions are filled with tight spaces that speak to the close-mindedness of the military and CIA officers therein. Figures come in and out of the shot with disturbing briskness and lack of clarity as if they were bodies without souls. The shots of the alien ships and even inside the ship, in contrast, are spacious, slow, and wide, speaking to the open-mindedness and creativity of the scientists and the aliens. Even the aliens ("Abbot and Costello") are afforded more personality and soul than the bodies of the humans. This is only one aspect of the *mise-en-scène*, but it is wonderfully expressive and thematically linked.

The cinematography follows the same project: the spaces inside the tents are dark and smoky, as if it were all shot on gray autumn days. The spaces in and around alien ship are airy and light, as if spring had arrived and landed in the middle of that autumn. Villeneuve has stated that he envisioned the film to have the feel of a rainy, dreary Tuesday morning— to his point, there is nothing special to such a day/time, yet the miraculous can explode through the mundane at any time. It is also a thoughtful and contemplative time; there is nothing sexy or alluring about a dreary Tuesday morning, so it begs for introspection and interiority.

Much of the cinematography is therefore soft focus vs. deep focus. Soft focus accounts for the past/future cuts and deep focus for the "present" with the aliens. Deep focus allows us to see both foreground and background in focus, and it is a more realistic mode of filming. We get to sit with the shot and the characters therein; the director is not telling us what to do or where to look—we can determine that for ourselves. Soft focus only allows us to see either background or foreground in focus; the other element is rendered as blurry. Soft focus is more of a formalist technique—we know, understand, and recognize that it is part of the cinematic apparatus—and it is less realistic. Each technique has its place, and each technique has its own resultant expression. Both fit into the overall plan.

The sound also expresses the thematic thrust of non-linearity and existence. Sylvain Bellemare, the supervising sound designer on the film saw his role as integral to the film and the overall feel and approach, noting "I approach sound as a character in the film." What a lovely sentiment! As I noted, the film won a (much-deserved) Oscar for Best Sound Editing. The sound in the film was meant to be natural, organic, and meaningful. One specific example of expressive sound speaks to these qualities: there is a distinct sound, or loud hum, around the outside of the spaceship shell. The idea is that it was meant to relate tension—in the sense that it was a tense barrier, a negotiation, and/or a challenge. The sound becomes a character in terms of how it holds its own unique specific attributes for a specific purpose. Bellemare also notes that most of the sound sources for the film were organic, once again speaking to the realism and authenticity of the overall look and feel of the film.

The final piece of this formalist film puzzle is the editing, which shocks and surprises us in following the same non-linear structure of the narrative as well as the rest of the cinematic aspects. "Non-linear time is a huge element of the film, and it is a huge element of the editing," says editor Joe Walker. For example, the cuts of the past/future are elliptical, and it forces us to think outside a linear narrative and outside a linear existence. The cuts are also fragmented, paralleling how we think and remember—in bits and pieces in our own minds. Then we fill in the gaps to create a whole narrative. Upon viewing it the second time, we better understand this elliptical nature and we are forced, trained if you will, to think as Louise thinks. The audience is therefore educated by the film as Louise is educated by the aliens.

As the kind of film that rewards for repeated viewings, herein lies the payoff: it speaks to a defining feature of humanity. As humans, we need to make sense of our existence, and to make sense of our existence, we need to tell stories. As we tell stories, whether it be our own history, family histories, or the tales of others, we have to fill in the gaps—the ellipses that are covered by the veil of memory or the fog of life. The elliptical nature of storytelling neatly rivals an elliptical nature of existence, rendering it non-linear. Whether it is the filmmaking process itself, a specific film, or our existence, that non-linearity offers a way out of simple causes and effects. In the process, *Arrival* offers us a glimpse of what it means to be human.

—Vincent Piturro

The Science of Communicating with Aliens

Ka Chun Yu

Since Georges Méliès' *A Trip to the Moon* in 1902, humankind has had countless cinematic encounters with aliens. In some rare cases, the aliens were inscrutable, with ordinary communication impossible (*2001: A Space*

Odyssey, 1968). But most of the time, communication with aliens was easy, since English turned out to be the lingua franca in Hollywood productions. In *The Day the Earth Stood Still* (1951), the lead extraterrestrial character appears on Earth able to converse in English (although having human characters memorizing and speaking words in his native tongue is crucial to the plot of the film). Even when intelligent simians speaking English should have been a giveaway to a key plot point of *Planet of the Apes* (1968), Charlton Heston and the audience take the Anglophone apes all in stride.

But even when aliens communicated in a language that was not English, they did so through modes that are familiar to us. Even when we hear alien languages with their own distinct vocabulary and grammar, they contain sounds audible to our ears, and are spoken in tones that human actors could articulate with their vocal chords; see for instance, the subtitled speech in *Star Wars* (1977) and *Star Trek: The Motion Picture* (1979). They could communicate via nonverbal means, but such alternate modes of exchange were still familiar to us, like the sign language and musical tones from *Close Encounters of the Third Kind* (1977), or the ethereal whale songs from *Star Trek IV: The Voyage Home* (1986). It is so rare to see alien communication that is *truly* alien, that when a film like *Arrival* attempts to do so, it is memorable how unlike human languages—how *alien*—an extraterrestrial language could be. Though they do speak, what is memorable about the heptapod language in the film is the ink patterns, sprayed squid-like, which then dissipate after lingering for a few seconds. And to add to its otherworldliness, the language does not unfold linearly the way that human languages are created, but is circular where past, present, and future can overlap.

If we ever have the opportunity to communicate with an intelligent alien species somewhere else in our galaxy, most scientists who have given this any thought do not think it will be done via any of the native tongues of either species. Instead, astronomers and scientists involved with the Search for Extra-Terrestrial Intelligence (SETI) argue that any common language for galactic discourse would be based on science and mathematics. If extraterrestrials are intelligent enough to build radio transmitters for bridging the interstellar divide, then they would be smart enough to understand the science underlying that technology.

The first attempt that humans made to communicate with intelligence elsewhere in the universe was based on this idea. The Pioneer 10 and 11 spacecraft were launched in 1972, the first of our robot explorers sent to reconnoiter with the outer planets of our solar system. They were hurled with enough velocity that they will eventually leave the gravitational embrace of the Sun and orbit the galaxy in interstellar space.

Attached to each of these spacecraft are gold-plated aluminum plaques designed by astronomers Carl Sagan and Frank Drake (Figure 1). The images of the nude man and woman might be the most eye-catching for most people, but the key to the message is the barbell-like symbol in the upper left corner. The two circles represent hydrogen, the simplest and the most common atom in the entire universe. A hydrogen atom consists of a heavy proton circled by a lighter electron. Both these particles have a quantum mechanical property known as "spin," which is only approximately like the more familiar spins that are associated with a toy top or a classroom globe. We will ignore what *exactly* spin is, but we only need to understand that the spins of the two particles in a hydrogen atom can be aligned with each other or against each other. When they are aligned, picture the electron and proton spinning in the same direction; that is, both moving clockwise, or both moving counterclockwise. For the spins that are oppositely aligned, one particle would rotate clockwise while the other counterclockwise. The barbell therefore represents hydrogen atoms, with the right one with proton and electron spins aligned, the left anti-aligned.

It turns out that when the spins are aligned, the hydrogen atom as a whole has slightly more energy than when the spins are anti-aligned. Furthermore, a hydrogen atom may start out with the spins aligned, but on average, every 10 million years or so, it will spontaneously transition to a state where the proton and electron spins are anti-aligned. When it does so, it loses energy in the form of a single photon of radiation that is emitted at a wavelength of 21 cm, or a frequency of 0.7 nanosecond (where 1 nanosecond is 1 billionth of a second). The probability for this to occur is rare, but given that there is so much hydrogen in the universe, this flip of the spins of lots of hydrogen atoms is something that human—and presumably alien—astronomers can easily detect, and are readily familiar with. The horizontal bar of the barbell represents the transition between the two different hydrogen atom states, as is the short vertical line below the bar.

Let us assume that alien salvagers find the Pioneer spacecraft far in the future. If they deduced the meaning of the two circles, then their next step in deciphering the plaque would be to notice that the two most common repeated elements are the vertical lines (|) and the horizontal hyphen-like bars (-). If they were to map these to a binary notation of 1s and 0s, then the 1 associated with the hydrogen atom transition could represent either a unit of length (21 cm) or unit of time (0.7 nanoseconds). The number 8 in binary to the right of the woman is meant to convey the height of the figures (8 times 21 cm or 5 feet 6 inches), which can be checked against the silhouette of the spacecraft, which is to-scale behind the human figures. The planets of our Solar System, including Earth from which the Pioneers originated, are arrayed along the bottom of the plaque, along with their relative distances from the Sun in binary notation.

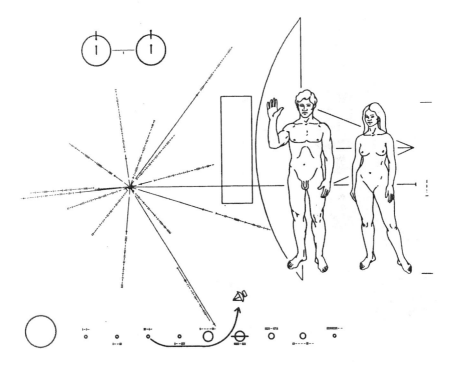

Figure 1: The Pioneer plaque containing information about where the spacecraft originated (NASA).

Finally, the dramatic-looking starburst dominating the left side of the graphic is a map back to our solar system, using the relative positions of nearby pulsars. Pulsars are rapidly spinning remnants of massive stars that have exploded in supernovae. The bright radio emission from the leftover core can wash over in our direction every time the pulsar rotates. These stars can start out spinning at hundreds of times per second, but they slow over time. Different pulsars will have different spin rates, and the map marks the spin periods and the directions of 14 pulsars relative to our Sun. Because the spin down occurs at known rates, an alien astronomer can determine when the probe was launched based on the pulsar rates at *its* current time versus the more rapid rates printed on the Pioneer plaque.

The Pioneer spacecraft were like messages in a bottle, cast out into the vast sea of space. They are small, so there is little hope that they would ever be encountered by a spacefaring race of aliens. A more reliable way of communicating would be to send out a radio transmission targeted at a star, and hope someone is listening. The first attempt at such a targeted transmission occurred on 16 November 1974, when the Arecibo Observatory in Puerto Rico was used to beam a nearly 3-minute-long burst of radio signals

20 The Science of Sci-Fi Cinema

Binary numbers 1-10

Atomic elements: Phosphorus, Oxygen, Nitrogen, Carbon, Hydrogen

Chemical formulas for sugars and bases in DNA

Shape of DNA with center vertical line showing number of nucleotides

Human height, human shape, and size of human population

Solar system

Arecibo telescope

Diameter of telescope

Chapter 1: *Arrival* 21

toward M13, a tightly bunched together cluster of 300,000 stars, located 22,000 light years away. The message was modulated by small changes in the radio frequency, which can be interpreted, Morse code–like, as a binary message. When the 1679 binary digits of the message are laid out in one of only two possible two-dimensional grids (Figure 2), a pictorial image appears. Again, the information embedded in the image encodes information about binary numbers, and progresses to even more sophisticated depictions of the elements important for life on Earth (hydrogen, carbon, nitrogen, oxygen, and phosphorus), as well as the DNA molecule responsible for our genetic code that is made up of these elements.

Since this initial attempt at METI—Messaging Extra Terrestrial Intelligence—we have discovered more than 4000 planets circling around stars elsewhere in our Galaxy. Based on some of our current surveys, it is estimated that up to a fifth of all stars have Earth-sized planets in the habitable zones of their parent stars. This means some 11 billion Earth-like worlds could be orbiting around Sun-like stars in our galaxy. Some researchers and groups are now interested in beaming radio messages to some of these nearby stars known to have planetary systems.

Ideas for messages have also gotten more sophisticated. In 1960, mathematician Hans Freudenthal published a book on *Lincos* (a contraction of *lingua cosmica*), his proposed language for communicating with extraterrestrial intelligence. It assumes messages are sent using radio signals. Different lengths of the signals are used to communicate phonemes, the fundamental elements of speech (as opposed to the letters used for a print language). The first message would start with simple mathematical concepts like counting to introduce natural numbers (1, 2, 3, 4, …), and then add additional symbols so that mathematical concepts like "100 > 10" can be communicated. From this foundation of symbolic logic, simple ideas are built up to allow complicated ideas to be expressed, including concepts from physics (such as measurements of length and time), and even abstract human behavior, and notions about life and death.

Lincos was the inspiration for Cosmic Call, a set of pictorial messages beamed from the radio dish at the Evpatoria Deep Space Center in Ukraine in 1999 and 2003. These communications, devised by astronomer Yvan Dutil and physicist Stéphane Dumas, start with simple mathematical expressions (Figure 3) and then expand from them to more advanced topics in mathematics (including geometry and algebraic equations); chemistry; physics; information about the Earth, Moon, and the rest of the Solar

Opposite: Figure 2: The Arecibo message decoded into a pictorial representation, with descriptions of the concepts that are to be conveyed (Arne Nordmann and Ka Chun Yu).

System; and even the biology of life on Earth (including information on the creators of the message). The messages are constructed from more than 100 symbols that are defined within each of the pictorial "pages." At the end, the very last part of the message asks for a reply back from any recipients of the message.

We cannot predict the exact form of the first message from extraterrestrials to us. They could be similar to the messages that we have concocted so far for communicating with aliens. But I bet they won't be anything like the human-alien engagements that we see popularized in movies and other media.

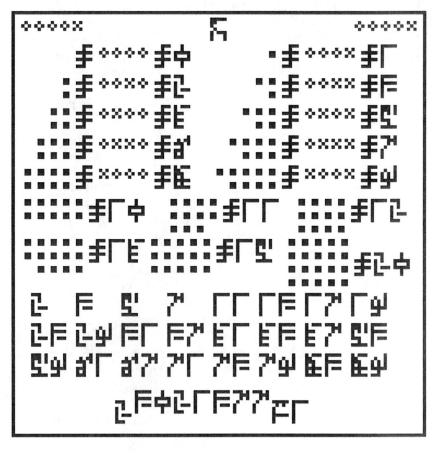

Figure 3: The beginning of the Cosmic Call message sent in 1999, showing an introduction to the numbers used later in the message (Yvan Dutil and Stéphane Dumas).

Arrival from a Linguist's Perspective

Andrew J. Pantos

The film *Arrival* and the short story on which it is based, "Story of Your Life" by Ted Chiang, raise some fascinating language-related questions. As a linguist, I was intrigued with the film's portrayal of the linguist and linguistic research methods, and with some of the underlying assumptions about language implied by the story. In this section of the chapter, I discuss some of my observations. Before I go any further, though, please let me emphasize that the thoughts expressed in this essay should not be viewed as criticisms or complaints. How could I complain about a film in which the hero is a linguist who saves the world? That's a great theme for any story. Nonetheless, given that the greatest concern of film directors is to create a good film that tells a compelling story, and not to accurately portray scientists and scientific methods (or trial attorneys, doctors, or anyone else) it's always interesting to see how film portrayals differ from the real world.

To review, *Arrival* portrays a linguist's attempts to communicate with visitors from outer space. These visitors are referred to as heptapods because they have seven tentacle-like legs (from the Greek *hepta* = "seven" and *pod* = "foot"). At the start of the film, the U.S. government recruits the linguist, Louise Banks, to help figure out how to communicate with the heptapods in order to find out the reason for their visit. The heptapods Dr. Banks encounters are confined to their spacecraft hovering above a field in Montana. It turns out that this is just one of twelve heptapod-manned ships hovering over various parts of the earth. Although the governments of the other host countries are also working towards the same goal of communicating with the heptapods, distrust among the various countries leads to a breakdown in scientific cooperation, and Dr. Banks is left to work alone on deciphering the alien language. In the end, Dr. Banks learns the language and determines that the aliens mean us no harm. Recommendations from the military for preemptive strikes against the aliens are abandoned in the nick of time, an armed confrontation is avoided, and the human race is saved.

The Portrayal of the Linguist

The film and the short story on which it is based perpetuate some stereotypes about linguists and linguistics that bear mention at the outset. First, linguists and translators are not the same thing. Translators deal in two (or more) known languages, both of which they presumably know very well. Translators do not figure out unknown languages and do not study language scientifically or theoretically. Translators are practitioners. The primary task that Dr. Banks faced was to figure out the language of the heptapods. Only after she figured out the language could she begin to translate between heptapodish and English. So, Dr. Banks's comment at the beginning of the film about testing the competence of her fellow linguist by asking him to translate a Sanskrit word is a bit odd. Knowing Sanskrit, or any other particular language, is not a requirement for being a linguist.

This mischaracterization of what linguists do is particularly interesting, given that later in the film it is made clear that the film's creators clearly understand that linguists analyze language (human language, in the real world) scientifically. This point is emphasized when Dr. Banks's astrophysicist colleague says in a shocked tone, "You analyze language mathematically!" and Dr. Banks gives him a dismissive look in response. In fact, linguists dedicate their careers to the scientific study of language, and there is an entire branch of linguistics dedicated to documenting undocumented languages. The astrophysicist's comment does, however, apparently accurately reflect a common thought among so-called hard scientists. Shortly after the film was released, a high-profile American astrophysicist criticized the choice of a linguist as the hero of the story, and he suggested that that a cryptographer should have been used to figure out the alien language, instead. On the contrary, a linguist is precisely the right person for this job. Cryptographers decipher codes that are based on existing, known languages. A cryptographer would not be helpful in this situation, where one of the languages is entirely unknown. A linguist fieldworker, who understands how to collect data and analyze language scientifically is exactly what is needed here.

The Portrayal of the Research Methods

The film's creators used three linguists as creative consultants on the film. Dr. Banks's methods for deciphering the heptapod's language are meant to represent the kinds of methods used by field linguists who document human languages. Just because linguists are hired as consultants, however, does not mean that their advice is always followed. After all, the purpose of this film was not to serve as a training film for future linguistic field workers, but to tell a compelling story. As such, some of the methods portrayed are a bit misleading.

Dr. Banks starts out by analyzing the physical properties of the sounds produced by the heptapods through acoustic measurements of the sound waves and spectrograms. This is an interesting choice. While acoustic measurements can tell us about the physical nature of the sounds produced, without associated meaning they cannot help in decoding the language. Dr. Banks could look for patterns in the physical properties of the sound waves and try to match those patterns to meaning, but it is unclear in the film exactly what she is trying to do with the acoustic measurements. The suggestion is that she is looking at the sounds waves alone to figure out the language. That would not be possible unless the sound waves were associated with specific meanings.

In fact, when they encounter a new (human) language, field linguists usually begin by noting the kinds of sounds produced by the speakers. In order to decipher the language, however, their focus quickly shifts to determining which sounds carry meaning in the language. In other words, linguists have to figure out which sounds are psychologically real to the speakers and which are not. Imagine a linguist encountering English speakers for the first time. The linguist hears one of the English speakers say "father" and another say "fatha." How does the linguist know whether these are two different words or the same word produced in two different ways? The linguist would note the difference, but until the linguist has figured out the sound-meaning pairing there would be no way of knowing that both productions refer to the same meaning. Now, if the one English speaker says "fatha" for both "father" and "farther," the linguist unfamiliar with English would have to figure out from sound-meaning pairings in context that two different meanings are associated with this same word. In both cases, a purely acoustic analysis of the speech produced would not help to decipher the language. The language cannot be understood until the sound-meaning pairings are determined.

Fairly early in her analysis, Dr. Banks gives up on trying to figure out the "spoken" form of the language (Heptapod A) and focuses her efforts on deciphering the "written" form (Heptapod B). I put the words "spoken" and "written" in quotes because it's not clear whether the heptapods are, in fact, speaking or writing. Our human concept of speaking requires the involvement of various aspects of our human speech mechanism (lungs, larynx, vocal tract articulators), all of which are based on human physiology. Who knows how the heptapods are generating the sounds they produce and whether we can really call that speaking? The same is true of the process of writing. Can we call what the heptapods are doing "writing"? In the film, it does not appear that the heptapods are manipulating any kind of tool or the surface where the symbols appear. Instead, the symbols are projected fully formed from the ends of the tentacles. Nonetheless, I'll use the terms here to facilitate the present discussion.

At this point in the film, the depiction of the methods used gets particularly problematic. Dr. Banks writes her name and the other scientist's names on a whiteboard and then shows the names to the heptapods, indicating whose name belongs to whom. If Dr. Banks's assumption is that the heptapods do not know English, this is not how most linguists would approach teaching them the language. How would the heptapods know what the written words refer to? We know it's her name, but how would the heptapods know whether she is referring to her name, her gender, her height, her age, her occupation, or any other characteristic of hers? What if the concept of individual names is alien to the aliens?

She continues to use this approach (called ostension) throughout. To teach the word "walking," she has her colleague walk. There is no guarantee that the heptapods, who do not themselves appear to have the physiology necessary to "walk," would associate that word with movement through space by placing one foot in front of the other repeatedly while standing erect. How would they distinguish this word from "move"? To confuse matters even more, Dr. Banks uses a derived form of the word ("walking") instead of the base form "walk." For human non–English speakers, this could be confusing. For aliens who do not use an alphabetic writing system, these—at least at first—might look like two entirely different words. Think about the difference between the word pairs "pit/pits" and "pit/spit." The addition of the "s" creates an entirely different word in the second pair, but the plural of the word in the first pair. Would non–English-speaking aliens understand the "s" the same way? In short, ostension is not always a great method of teaching a language, and the use of derived forms of words, particularly at first, is potentially confusing. Interestingly, Dr. Banks herself makes this point in the film with her anecdote about the confusion surrounding the meaning of the word *kangaroo* when it was first borrowed from the indigenous people living in Australia.

Another interesting choice is the depiction in the film of the exclusive use of intransitive verbs in her ostensive research. Intransitive verbs are verbs that have no direct object. For example, "arrive" is an intransitive verb because we can't arrive *something*. By the same token, "kick" is a transitive verb because we can kick something. Early on, Dr. Banks discovers that the heptapods' writing system, Heptapod B, is non-linear and order-independent. That, of course, is very different from English writing, which is linear and order-dependent. In fact, word order is particularly important in a language like English. The two sentences "The cat chased the mouse" and "The mouse chased the cat" mean two different things solely because the noun phrases "the cat" and "the mouse" have been swapped. There is no difference in the form of the words "the cat" or "the mouse" that would indicate which animal is chasing which. Word order alone determines the difference. Word order should be of paramount importance in trying to teach English to speakers of a language that is not order-dependent. In order to stress the importance of word order in English, both transitive and intransitive verbs would have to be used.

Similarly, an odd choice was made in the sentence that Dr. Banks claims to be working towards translating into Heptapod B. She writes the sentence "What is your purpose for coming here?" on the whiteboard to illustrate how difficult it is to convey all the aspects of that sentence in another language. True. So, why not just translate the sentence "Why are you here?" It is unclear why Dr. Banks chose the more complicated sentence to translate when the simpler sentence would have conveyed the exact same meaning. A linguist would go for the simpler sentence.

The Portrayal of the Alien Language

Obviously, everything we know about language is based on what we have studied about human language. So, are humans even capable of conceptualizing an alien language? Human conceptions about language, including alien languages, use human language as a starting point. A number of science fiction and fantasy writers have created their own languages for their unreal characters: Klingon, Dothraki, and Avatar, are just three examples. Historically, one of two approaches has been taking in creating these languages. Tolkien used influences from a variety of ancient and modern languages as a basis for his Elvish languages. Finnish, in particular, influenced the grammar of Quenya, one of the Elvish languages, and Celtic languages influenced Sendarin, another of the Elvish languages. Tolkien's languages were intended to seem other-worldly, but in fact were based on qualities of actual human languages.

Another approach is to look at the typical, most frequently encountered characteristics of human languages and create a language that is as different from known human languages as possible by using a combination of the rarest characteristics. If, for example, most languages have a word order that is Subject-Object-Verb (e.g., Japanese) or Subject-Verb-Object (e.g., English), then the creator would choose a different, less-common word order (e.g., Object-Verb-Subject) for his or her fictional language. If few languages allow certain sound combinations or tend to omit certain sounds from their phonemic (sound) inventories, those will be the sounds that are used in the created language. To create an alien language, then, the goal appears to be to create the most foreign sounding language possible to the largest number of people in order to convey the sense of "alien." For example, in *Avatar* (2009), the Na'vi language, in terms of both its sound system and its syntax, was designed to sound as different from human languages as possible by incorporating the least common syntactic structures and the least common sounds in human language. These approaches seem to work convincingly where the language is just one of many aspects of the characters that are selected to create the alien depiction, and where the language itself is not the focus of the story.

Arrival is different, however. The language of the heptapods is central to the plot of the film (and to the short story on which it is based). This is where a fundamental question arises. How can a language that is independent of our human conceptualization of language be created, when creating a language that consciously either does or does not have characteristics of human language necessarily means creating a language from the perspective of human language? In either case, one can argue, the language is based on human language.

If we set aside that concern and focus on the language created, we see that Heptapod B is characterized as being non-linear. The circular symbols are called ideograms in the film and each symbol is meant to convey a complete thought. Each ideogram has no obvious beginning point and the ideograms are presented in no obvious order. Once Dr. Banks figures all this out, she is able to communicate with the heptapods. In addition, once she figures out the non-linear aspect of Heptapod B, she is able to experience time in a non-linear fashion, living through the past, present, and future simultaneously.

While the concept of circular ideograms is clever, the notion that they could have no beginning point and still be intelligible is difficult to maintain, particularly for an English speaker, a language for which word order is critical to sentence meaning. If I start the sentence "the cat chased the mouse" with "the mouse," I'd come up with "the mouse the cat chased," which could be interpreted as either the cat chasing the mouse or the mouse chasing the cat. I can't present the words in the sentence in a circular fashion with no designation for the starting point and hope that the reader will understand with I'm saying. Furthermore, arranging all the sentences of a paragraph in no particular order on a screen and asking someone to figure out the meaning of the paragraph. A number of interpretations would be possible. Determining meaning, particularly in the relatively short amount of time depicted in the movie, would be impossible. If, on the other hand, the ideograms are not composed of words, but of entire concepts, we have the problem of how to pictorially depict abstract concepts. How would "We desire the international cooperation of all human linguists" be depicted in an ideogram? It's an interesting question, and perhaps impossible for a human to produce. That may be why the ideograms are never deciphered for the audience. Another question that is left open is how any depictions could possibly work across alien (literally, alien) cultures.

As I mentioned, the film and story suggest that learning Heptapod B enables Dr. Banks to experience time differently. Linguistic Determinism, the idea that language determines thought (the so-called "Sapir-Whorf Hypothesis") has been debated and tested for decades. While its most extreme interpretation—that we can conceptualize the world only in a way that our language permits—is no longer seriously argued, the softer version that maintains that language can affect the way that we see the world is. There is research that has shown such effects, particularly with regard to color terms and grammatical gender markings. If my native language has only five color terms, then, according to the theory, I categorize the colors that I see in terms of those five colors. If my language grammatically categorizes a "bridge" as a feminine noun, then I will tend to use adjectives that describe the structure as having stereotypically feminine characteristics (e.g., elegant, pretty, slender). If my language categorizes the word "bridge" as a masculine noun, then stereotypically masculine characteristics tend to be attributed to the bridge (e.g., sturdy, strong, towering). Accordingly, we can say that language can have an effect on how I see the world.

Without having to debate the merits of the theory in general, however, we can argue that its application here to second language acquisition is a leap. This is even more questionable in the situation in Arrival. As depicted in the film, Dr. Banks has studied Heptapodish B for only a few weeks or months (it's unclear just how long she studied the language before deciphering it), and only this written form. No linguist I know would maintain that learning the written form of a foreign language for a few months is going to rewire your brain. Learning a foreign language is not the same as acquiring your first language. As we acquire a second language, we are constantly comparing it to our first language. A second language is always acquired in terms of the first language. The acquisition of the second language by adults is not integrated into our development process as our first language is. Learning a second language may mitigate the constraints of our first language to some extent, but is unlikely to rewire our brains entirely, and certainly not if we have studied the foreign language for only a brief period of time.

One final linguistic aspect of the film and story that bears mention is an interesting explanation of the language lessons that is circulating on social media. This argument posits that the heptapods did not need to learn English, because the heptapods had already learned English on a previous visit. (I presume they learned other human languages, as well?) Anyway, according to this theory, the real purpose of the visit was to get the international community to work together cooperatively on one project, namely figuring out the purpose of the intergalactic visit, in order to teach us to love one another. (The fact that Dr. Banks learned Heptapod B and was able to see time non-linearly was a side benefit, according to this view.) If this is the case, the heptapods are coy devils, indeed. It's an interesting thought, and one that completely de-emphasizes the role of language. Instead of making the breaking down of the language barrier as essential to figuring out the purpose of the visit, it becomes one of a number of tasks that the heptapods could have set in motion to stimulate international cooperation. It also leaves open the question of what would have happened had Heptapod B not been deciphered, and the military option had been pursued. I, for one, have to believe that Dr. Banks' efforts were necessary and that the linguist really did save the world.

Chapter 2: *Interstellar*

The spaceship *Endurance* approaching the black hole in *Interstellar* (Paramount Pictures, 2014).

Christopher Nolan is one of the best directors working today—a current *auteur*. His second feature *Memento* (2000) made him an A-list director, and he has since made a seemingly endless string of great films/hits, including the last Batman trilogy, *Insomnia* (2002), *The Prestige* (2006), and *Inception* (2010). Originally a vehicle for Steven Spielberg, *Interstellar* was handed off to Nolan with a $165 million budget, a cast of A-list stars, and the universe as his canvas. His brother Jonathan, himself a noted writer (HBOs *Westworld*), helped to re-write the script, and noted astrophysicist Kip Thorne was the science advisor on the film. The A-list cast included starring roles by Matthew McConaughey, Anne Hathaway, and Jessica Chastain, as well as meaty supporting roles by Matt Damon, Michael Caine,

John Lithgow, Ellen Burstyn, David Gyasi, and Wes Bentley among others. The film was big-budget studio sci-fi for grown-ups from the beginning, and it delivered.

It is a sweeping, epic story: in the near future circa 2070, blight has struck the food supply while widespread famine and drought pervade the Earth. Most scientists agree that end times are near, and they need to take drastic measures. Around the same time, the scientists find an inexplicably placed wormhole near Saturn, which they use as a portal to other galaxies. Several years earlier (before we enter the film), the underfunded and nearly-defunct NASA had sent out explorers to all of the planets deemed possibly hospitable, and the time has come to send a follow-up team to those planets/explorers and begin the process of continuing the human race. It is either a suicide mission to the unknown or the beginning of a new era. Joseph "Coop" Cooper (McConaughey), a farmer and former engineer/pilot in the middle of what looks like a throwback to Dust Bowl America (replete with flying drones and robotic tractors), helms the mission to re-trace the steps of the explorers and see if there is, in fact, a suitable replacement for Earth. In the process, he leaves his daughter behind.

There is a lot going in the film, with its start on Earth (dusty farms and towns) and its finish in another galaxy (planets dominated solely by water or ice) next to a black hole, including trips back and forth in-between those two points throughout. The heart of the film, however, is the relationship between Coop and his daughter Murph, a relationship that takes on extraordinary challenges over decades and across the universe. Murph is represented in three different stages by three different actors, and the relationship that spans galaxies pushes current astrophysics to its limits and beyond. In the first iteration of Murph, ten years old and played by a wonderful Mackenzie Foy, she is a child prodigy living on the family farm with her father, brother, and grandfather. She causes trouble at school for bringing in books telling the story of the Moon landing, evidently a no-no. Such stories have been corrected in this near-future world, the teacher tells Coop, because the real story is about how the moon landings were faked to bankrupt the Soviets—the point being that the moon landing is irrelevant in this world of famine. The principal starkly drives home the point as he tells Cooper about his son and his son's future: "we don't need more engineers; we need more food." We then see Murph at thirty-five (Chastain), now a renowned scientist trying to figure out the magic formula for saving humans. She is brilliant but bitter: she believes her father has abandoned her and never intended to return. This beautiful but difficult relationship is the center around which the rest of the magnificent narrative and visuals revolve.

The film was shot mainly in IMAX 70 mm and anamorphic 35 mm, the shoot lasted 5 months, and it spanned three countries. Overall, the finished piece of art is a formalist gem, including challenging *mise-en-scène*, cinematography, editing, and sound. Everything works together to tell the story, invest the film with specific style, and point toward multiple themes. The production design and visual design are stunning, and the mathematical representation of a black hole (designed by Thorne) was so good that it is now used as a teaching tool for astrophysics students at Cal-Poly. It was critically well received, and it did moderate business at the U.S. box-office with a much better international showing.

With the big budget, a director at the height of his powers, a stellar cast, a basis in real science, and astounding special effects, the film still remains profoundly human. The relationship between Murph (Chastain as Murph in her '30s) and her father (McConaughey) may be the central force, but it is by no means the only one. From the start of the film to the last frame, humanity and its quest to remain relevant are foregrounded among all of the trappings of Hollywood. Much like *Arrival*, *Interstellar* wonderfully address the central question of science fiction in unique and powerful ways: "What does it mean to be human?" We look first to all of the filmic aspects to get us there.

The *mise-en-scène*, cinematography, editing, and sound all point the way toward a central thematic project in the film; for such a huge project infused with mind-boggling special effects, that central project is one of realism. The realism manifests itself in many ways—small and large— and speaks to the urgent concerns at the center of the film as well as its presentation and effect. We are able to see the past, present, and future in terms of that realism, with the net effect being something tangible and visceral. The verisimilitude also speaks to an urgency in the film that reflects a larger trend in recent science fiction cinema: the future is close and the issues are laid bare, at our doorsteps. We are no longer allowed to pawn off our current and near-future problems onto a faraway future that makes us feel safe and comfortable. Science fiction has always been great at pointing out current issues while projecting them onto a future plane; *Alien* (1979), for example, spoke to environmental degradation on Earth but set the film several hundred years in the future. *The Matrix* (1999) is set in 2199. With more recent sci-fi, including *Interstellar*, that is set in the near future, the result is that we are no longer given that safe distance from which to view the destruction. Recent sci-fi has closed the temporal distance that used to offer a false sense of safety and now speaks to our current situation—a world in which we can't ignore crop disasters or climate change. The future is now.

Thus, the realism. Nolan has stated that he was interested in realism for the sake of clarity and sense of purpose. For example, he wanted the opening sequence on the farm to be appealing, simple, and beautiful, yet still on the verge of disaster. If it wasn't appealing, Nolan stated, then it wouldn't be difficult to leave. It is not a dystopia but rather a dying utopia. Nolan wanted to eliminate trying to figure out what the future would look like and made it look like today: "Who cares what their trousers would look like," he mused. That beauty and simplicity extended to the filming itself: on location in Canada, the crew defied nature in interesting and unexpected ways, with remarkable results. This opening sequence also does what I see as an important element of science fiction: it grounds the film in Earth. Even though the narrative crux of the film revolves around the degradation of Earth, it still makes clear that Earth is not only a place that has been ruined, but it is something that needs to be *found again*. Earth is thus very much the center of the search. This focus on realistic aspects of the *mise-en-scène*, from the dusty, dying surface of the planet, to its lonely beauty in the vastness of space, supplies the film with a mix of wonder and horror that has seeped into the world consciousness in our time. It is real.

The realism extends to the smallest details in the film as well. Nolan wanted the costumes to be authentic in terms of their usefulness and utility. Space suits were based on updated versions of NASA spacesuits—more classic than futuristic. Everything on the suit has a purpose; nothing is decorative. Suits weighed 30–35 lbs. with a cooling mechanism in them, similar to real astronauts. In so many science-fiction films, the suits are ludicrous representations of the real thing that wouldn't last a minute in real space; here, they are modeled after real NASA suits and the actors wore the real thing, including working microphones inside the suits.

That attention to realistic details suffuses into the sets as well, including not only the farm and the farmhouse as mentioned before, but to all of the spaceships as well. Modeled after the NASA Space Shuttle and the International Space Station, Nolan wanted everything to be utilitarian, and therefore nothing is superfluous. The handles inside the ships, for example, are placed so the astronauts can pull themselves around easily and logically; the control panels move and adjust so they have multiple uses. While many other sci-fi films would include buttons and knobs that don't do anything, Nolan tried to design everything with a purpose.

The locations even follow this logic of realism and CGI was only used to enhance the actual location shoots. The farm scenes were shot on location in Canada, one of the planets was shot on location in Iceland, and the other planet was shot on a glacier in another part of Iceland. On the Icelandic glacier subbing for Mann's planet, the sets were built right onto the ice itself and had to be adjusted every day due to shifts in the ice and the brutal conditions—ice, snow, and wind. On Miller's planet, which is entirely covered in water, the crew found an area with shallow water for miles. The cast spent days in wetsuits with water up to their knees. All reported a cold, taxing few days of shooting.

The Canadian location for the farm sequences speaks to the film's yearning for authenticity and thematic depth in particular. The location outside of Alberta, Canada is not what you might think of when picturing the American Midwest—its rolling hills leading up to a mountain range in the background reminds more of a Colorado or Montana landscape than it does Iowa or Kansas. Seeing miles of corn plopped down into this landscape screams of disjunction. And that is exactly the point—the images were meant to be evocative of something that is not necessarily natural, but rather, forced; it is a venture that is pushing the limits of human knowledge and experience while pushing the boundaries of the Earth's natural tendencies. In short, it feels like a last-ditch effort, a gambit one endeavors when all else has failed. The physical space and the images therein thus speak to a planet and civilization in decline: a world in which all other possibilities have been exhausted and desperate measures are all that remain. The simplicity of the farm and its surroundings belie the fact that the Earth is dying, including the people along with it. The opening sequence therefore speaks to the holistic project of the film and its vision: we are failing our planet and we must take drastic measures to save it and/or ourselves. It is certainly an environmental message in an age of climate crisis and one that suggests we look not only inward for relief, but to the stars as well. And that message speaks loudly through the *mise-en-scène*.

Yet even in this near-future milieu of decay, the production itself finds hope and freshness in the character of human ingenuity. The actual location—outside Alberta, Canada, and a place that has been the setting of many an American Western—was not suited for growing corn. The production built the house of its dreams in this valley, but they were told that corn would not grow there. CGI was an option, but Nolan opted for realism and the production planted its own acres of corn on the land around the house. To everyone's surprise, it worked. The corn thrived, the images were stunning, and they even harvested most of what they planted with the profits helping to offset the costs of the location! The monster science fiction film with the ridiculous budget scored their first profits on the film (long before it was released) by planting, growing, and selling corn. All in a location that was not supposed to support it. Fitting.

That realism of the opening sequences is buttressed by the intercutting of interviews with people who seem to be speaking of an era reminiscent of the 1930s American Dust Bowl. When we see the awesome storms of that portion of the film, it certainly has the aesthetic of the dust bowl and was actually based on a real storm from 1935 Texas. The production team used a biodegradable additive and blew mountains of it around the town of Fort

McLeod (outside Calgary) to create a realistic version of such a storm. The cast, crew, and the town itself took days to clean up after the shooting, all in the service of verisimilitude. The interviews—themselves actual recorded Dust Bowl remembrances from survivors of the era, are in the Ken Burns style of wistful and nostalgic referentiality—seem out of place until we get to the end of the film and understand their significance and their placement. In the grand scheme of the film, they come from the future: the elderly Murph and other farmers/inhabitants of this era are speaking of what it was like to live in these last days on Earth. The interviews would be part of a historical museum-type piece established on the Cooper space station conceived and built by Murph with data transmitted to her by her father, in the future. The interviews are plopped down into the beginning of the film, however, given to us by Nolan in the same gravitational time-slip-maneuver whereby Cooper gives his daughter the data to save civilization. If it all seems so complicated as you read it, well, you must see the film. But the film's *ethos* is firmly established—in every way—in this opening.

The cinematography also follows this project of realism and authenticity—as much as it can. And this probably a good time for a disclaimer: it is still *a movie*, and even film realism only approximates the reality of any place or time in its lens. As co-writer Jonathan Nolan states, "At one point we are standing in a fake dust storm next to a fake house surrounded by fake corn in a fake Midwestern location." But it looks real, and it felt real to the cast in particular, who were able to act next to and off of real locations, real props, and real subjects. For example, the chase through the corn to find the wayward drone was shot in the same location as the action in the diegesis—the corn fields around the house. Coop and his kids see the drone and then chase it, plowing through 7-foot-high corn stalks to do so. To get the effect, the crew used a model airplane that was actually flying over the truck, someone was actually driving the truck through the cornfield, and everything was directed by a helicopter pilot flying above them while the crew shot inside the truck and from outside the helicopter. There are effects involved, to be sure, but the majority of the sequence was shot right there, on location, and it actually happened. As the CGI director noted about the film, "When I read the script, I thought we would have to do most of the work supplemented by the shooting, but it was the other way around." The actual location, perspective, and moving camera of the cinematography place us into the action, make us feel as if we are there, or at least, that *it happened*. That is the essence of film realism.

The same holds true for some of the space sequences, which were shot mostly on location with only minimal effects intertwined with the real footage. As noted, the water planet was shot on location in Iceland, as was the ice planet. Matthew McConaughey and Matt Damon actually fought on the ice. The base literally blew up (in one take!) on the ice. The windstorm actually came through during filming and became part of the sequence. The crew had to continually tweak the sets as the ice moved several inches daily; bolts that held down the structures one day would be uprooted the next. The wind was so bad at one point that the crew had to retreat to the hotel and wait it out. Unhappy that they were sitting idle, Nolan set up shop in the parking lot of the hotel, shooting close-ups and reaction shots so they wouldn't waste any time. The conditions were always conducive to actors feeling as though they were in a different place and time, however, in real conditions. As good as these actors are, you can't fake being actually cold or actually standing in freezing water up to your knees. Again, the viewer is given a sense that we are there and this is real.

The lighting inside the various ships was also designed with realism and function in mind: the lights emanated from inside the spaces of the ship (the actual lights that would be on such a ship) but still gave cinematographer Hoyte Van Hoytema (*Dunkirk, Let the Right One In, Her*) the lighting he needed to get the shot. Even though the ships were constructed on sets, Van Hoytema brought his camera into those sets, in tight spaces, with only the ambient lighting they built into the ships. It is quite remarkable when considering how they could have lit the sets otherwise, standard practice for any Hollywood film. Instead, they opted for realism.

The editing follows this same project of grounding the film in a realism that allows the viewer to suture themselves into the film. The cross-cutting sequence of Cooper on Mann's planet while Murph and brother are back on Earth, on the farm, is a great example as it follows this scheme but also highlights the film's wonderful marriage of form and substance. We are given realistic settings in both locations—the Icelandic glacier standing in for Mann's planet and farm in Alberta. Thematically, both sequences convey a struggle for survival, a struggle to save themselves, and a yearning for family. Cooper fights Mann (instigating a survival instinct that Mann narrates to him) as Murph fights her brother to save him, his wife, and his son. The effect of the cross-cutting is thematic and realistic on both ends. Consider this entire sequence as form meets content: the *mise-en-scène* is realistic and authentic, including the blistering cold of the planet and the dust storms/fires of the town on Earth; the cinematography uses mostly handheld camera to give a documentary-feel to the Earth sequence as well as the fight on the planet; and the editing connects the two plots together through the cross-cutting, pointing to the yearning for connectivity that grounds the film thematically.

That editing also has a physical property to it that mirrors the events of the film. As the sequence moves on and Coop brings the ship to the brink of the Black Hole and jettisons Brand off to the lone remaining promising planet, we continue to cut back and forth between Coop entering the Black Hole and Murph examining her old room for clues. We are dealing with ruptures (or rips, or whatever we should call them…) in time and space at this point. Coop is entering the sixth dimension inside the Black Hole and about to send back the messages to young Murph, who believes it is her ghost. At the same time, older Murph is examining the room and finds the watch that Coop codes with the information she needs to solve the equation back on Earth. The editing ties all of this space and time together into a coherent story so that we, the viewer, are able to follow along and see how these three different places and times all fit into the larger narrative. The overall film and the cross-cutting of the editing are no longer bound by the "normal" laws of cross-cutting—parallel action showing two different things happening in different locations at the same time—but rather, now we see three different things happening in three different locations in space at time, at presumably the same time. In other words, the editing wonderfully mirrors the physics of the film.

The music is the final piece to this intricate puzzle of a film. It follows along with the project of realism while adding depth and breadth to the entire construct. Nolan had very specific ideas about the music, including the feel and the progression in particular. He told composer Hans Zimmer that he wanted organic sounds, and he wanted the music to change along with the physical movement away from Earth—the music close to Earth was very much melodic and *of the Earth*, while the music as they moved further away would become increasingly warped. Nolan also wished to escape the limits of what we have seen in previous science fiction: "It was very important to me that the music not pay attention to the genre," said Nolan. He gave Zimmer a task: write one page about the relationship between a father and his son, but he didn't tell him what the movie was about. Zimmer himself has a son, so he wrote it about his own relationship to his own son. That simple organ piece Zimmer wrote would become the leitmotif of the score and the basis for the remainder of all the music. They even recorded the organ in the Temple Church in London so they could get real sounds in a real place. Once again, Nolan was striving for an organic atmosphere with heart and humanity.

Further, the sound and music corrode as you get further away from Earth, and the instruments are meant to remind you of "everything you can lose." This once again plays right into a prominent subject of the film, connection, with the theme being that humans are constantly searching for connection: on a personal level (like a parent to a child in Coop and Murph) or even on a larger level, such our connection to the Cosmos. We are constantly searching for those meaningful connections in our lives in many ways, and we also seek to connect to the universe—be it a connection to other beings or a metaphysical connection. The music thus rounds out this central theme of the film and it all fits squarely into the larger question of science fiction: "what does it mean to be human?" *Interstellar* answers this question with wonder, poetry, and humanity.

—Vincent Piturro

How Easy Is *Interstellar* Colonization?

Ka Chun Yu

In Christopher Nolan's *Interstellar*, crops are failing all over Earth because of blight in the near future. Matthew McConaughey's character becomes involved in a NASA mission to look for other habitable planets for humans to escape to as Earth becomes increasingly uninhabitable. One can read this plot device as a reflection of our modern anxieties about how climate change and environmental distress may force us to look for opportunities out in space. Public figures like Stephen Hawking have advocated for humans eventually spreading out to the stars, while Elon Musk has stated he wants to colonize Mars.

The idea of colonizing other planets long predates its present-day promotion, but it has been a venerable theme in science fiction. Before the Space Age, it was common for American writers to imagine that colonizing nearby planets in our solar system would be a futuristic extension of how Europeans colonized the western hemisphere. But starting in the mid–20th century, our Earth-based remote sensing technologies and our first interplanetary space probes began to reveal what these worlds were really like. We learned that the surfaces of Mars and Venus were much harsher than we had imagined, both extremely dry, as well as unbearably cold or hot. There was no chance that intelligent aliens ever appeared on these worlds. At best, we could hope that microbial life had evolved when Mars was warmer and wetter billions of years in the past, and a few of these single-celled organisms could still be tenaciously holding on deep underground, waiting to be discovered.

But even given its inhospitable surface, Mars beckons to those who feel humanity should not be bound to just one planet. Given what we know today, how difficult would it be for us, not just to visit with a small team of explorers, but to plant large human colonies on other planets? What

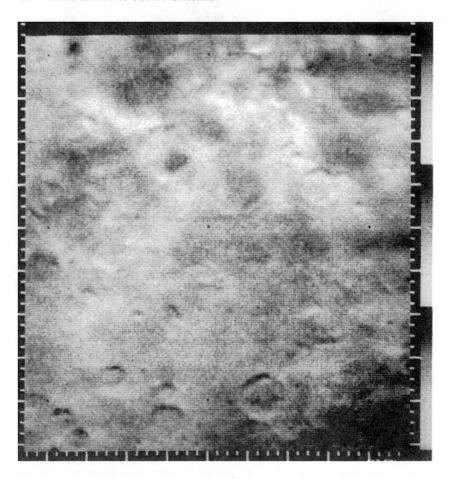

The first image beamed back from the Mariner 4 spacecraft in 1965 of Mars' surface shows a surface pockmarked by craters, revealing a world very much unlike the one built up in people's imaginations up until then (NASA).

attributes of a planet make it suitable for life as we know it? And even if a world like Mars is harsh and uninviting today, how difficult would it be to terraform the surface to make it possible for humans to live on the surface without wearing spacesuits? And following the lead of *Interstellar*, how hard would it be to colonize Earth-like planets elsewhere in the universe, assuming that the problem of transportation is already solved for us, such as via the wormholes seen in the film?

First let's make a list of requirements of what is necessary for life as we know it. We need oxygen in the air for animals to breathe, and carbon dioxide for plants to grow. Nitrogen must also be available in the environment for bacteria to "fix" it into a form that can be readily used by plants (and ingested by the animals that eat the plants). Water is a must for all life, and the surface environment must be in a temperature and pressure range conducive for it to exist in an accessible liquid form. Finally life must also have some protection from damaging solar ultraviolet (UV) radiation as well as high energy particles in the form of cosmic rays and solar wind particles.

Since the formation of the solar system, Earth has evolved over the last 4.5 billion years so that multiple geophysical and biological systems are involved to supporting life. Plants and cyanobacteria grow by taking in water, carbon dioxide (CO_2), and sunlight, and converting it to biomass, The oxygen (O_2) that makes up 21 percent of the atmosphere, and which animals breathe in, is released as a by-product of the photosynthesis reaction. The 78 percent of Earth's atmosphere made up of nitrogen is tapped by bacteria and converted into a form available for use by plants. Since they are also the source for food and resources like building materials, clothing, and pharmaceuticals for humans, future self-sufficient colonies will have to plant and keep alive a diverse range of plants.

The trace amount of heat trapping gases carbon dioxide or CO_2 (about 400 parts per million and growing, because of human activities) and water vapor in our atmosphere prevents surface heat from radiating completely back into space. Without these, the ground of our world would be, on average, close to freezing. Water is found in abundance on the Earth's surface in liquid form, although there is enough variability in surface conditions for it to exist in gaseous and solid forms.

A small fraction of the oxygen molecules in the atmosphere are further converted into ozone (O_3), which in the upper atmosphere, helps shield some of the Sun's dangerous ultraviolet (UV) radiation. Even the Earth's magnetic field, which in addition to letting us navigate with compasses, has an additional role in protecting us from harmful radiation. This magnetic field can deflect high velocity charged particles from deep space, some of which are re-funneled towards the polar regions, where they slam into the upper atmosphere to create aurorae. By lowering the rate of energetic particles impacting our atmosphere, the magnetic field also lowers the amount of gas that gets stripped and lost into space.

Living on Mars

Mars is often spoken of as the world most similar to Earth in our solar system.[1] But in many respects, it is dissimilar in many of the critical ways involving habitability. The Martian atmosphere is much thinner, with the pressure at the surface only 0.6 percent that of Earth's surface. Even though the atmosphere is 96 percent CO_2, the layer of air is just not thick enough to provide much of a greenhouse effect to warm the surface. As a result, the average temperature over its surface is about -60 deg C or -76 deg F.[2] The low pressure of the Martian atmosphere also makes it impossible for liquid water to exist on the surface. Instead of freezing, any water poured onto the surface of Mars simply sublimates, or turns directly into a gas. The lack of oxygen means that there is no ozone to shield the surface from the Sun's damaging UV rays. And the absence of an appreciable magnetic field also means there is no shielding from charged cosmic rays and solar wind particles. The surface of the world is under continual radiation bombardment. If there is microbial life on Mars, it is unlikely to survive in the topsoil, but would have to be found deeper. Similarly, astronauts visiting from Earth would be exposed to greater amounts of damaging radiation on the surface of Mars than they would on Earth. Habitats for human explorers may have to be buried underground, where a top layer of rock and dirt can help protect the inhabitants from the flux of cosmic radiation.

Despite such hostile conditions, one could imagine sending small teams of astronauts to explore Mars and live "off the land." In order to do so, they would have to be able to grow food in Martian soil, instead of hauling in all the supplies they would need for a multi-year mission. Luckily observations from our robotic explorers at Mars suggest the planet has substantial amounts of water in the polar caps and underneath its surface. If all of the polar cap water was melted, it would cover the surface of Mars to a depth of 22 meters (72 feet). Ice that has been indirectly detected underground could provide add another 12 meters (39 feet) to that global ocean. It would take an immense engineering effort and considerable expense of energy, but there is no technical reason why astronauts would not be able to mine and melt ice from underground reservoirs, or transport meltwater from the polar caps.

Given sufficient water, concentrated CO_2 from the atmosphere, and imported fertilizer which would provide fixed nitrogen, could our future Mars explorers grow crops like Matt Damon's character from the movie version of *The Martian*? Unfortunately, real Martian soil may not be so conducive to agriculture. Over the last decade, our robotic landers and orbiters at Mars have discovered that the reactive chemical perchlorate appears to be globally distributed on the surface.[3] Its concentration is only 0.5 percent, but it is enough to be poisonous to many microorganisms. On Earth, plants grown in soil contaminated at lower levels of perchlorate can accumulate the chemical in its foliage. People who ingest such vegetation would have their iodine uptake by the thyroid gland affected, which can affect metabolism and growth. In order to grow any food, Martian soil will have to be cleansed of all perchlorate. Scientists Yoji Ishikawa, Takaya Ohkita, and Yoji Amemiya have estimated that at least 100 square meters (1080 square feet) of land is needed per person underneath domed greenhouses in order to support a human colony. However as an insurance policy against accidental breaches of the greenhouses which would immediately flash-freeze and kill your crops, they suggest 1200 square meters (12,900 square feet) per person, which would mean just over three football fields of land would be needed for a small colony of 15 people. Assuming at least a foot of topsoil is necessary, roughly 20,000 tons of Martian soil would have to be processed to remove all perchlorate. At the low end, just under 2000 tons of Martian soil would have to be cleansed.

In *The Martian*, the main character has to grow potatoes in Martian soil because he was unexpectedly stranded on Mars with not enough food to eat before he could be rescued. But future missions that rely on growing crops on site will likely use hydroponics, where plants are suspended, with only their roots exposed to concentrated solutions of water and nutrients. Current research sponsored by NASA shows that about 50 square meters (540 square feet) of greenhouses is needed per person when using such a technique. Thus even accounting for insurance against technology failures and accidental greenhouse breeches, doubling this number to 100 square meters (1080 square feet) per person means that our hypothetical 15 person colony on Mars will need just under a third of a football field of greenhouses. However, studies also show that very strong lighting is needed to give crop yields high enough to support the amount of calories the crew would need. At Mars' distance from the Sun, natural sunlight will not be strong enough. The greenhouses will need to have mirrors to concentrate the sunlight, or high intensity artificial lights need to be installed. Considering that global Martian dust storms can turn the Martian day into night for months at a time, artificial lighting will need to be in reserve even if the greenhouses use natural lighting most of the time.

Hydroponic systems have many advantages, including allowing precise amounts of water and nutrients to be sent to the plants, which helps minimize water waste. But because the plant roots are not resting in soil, it is important that the mechanical systems run smoothly. Crew members have to be ready to fix or replace an electric pump that has failed, and plug up leaks or remove clogs in the plumbing. Breakdowns in the hydroponics support infrastructure can be deadly for crops whose suspended roots can dry out, compared to vegetation planted in soil that holds onto moisture.

Therefore, a Martian base that is self-sufficient in food will need ample power to run lights, heaters, fans, and pumps to support agriculture. Because mechanical parts wear down over time, a large supply of spare parts needs to be on hand to ensure that the inevitable breakdowns over time do not lead to catastrophic crop losses. The mission planners need to ensure enough supplies are brought on the initial mission or can be re-supplied over time. There need to be not only gardeners and farmers, but also engineers to maintain the mechanical systems that sustain the agriculture.

Making Mars Livable

Making the Martian surface habitable for many more people would require it to become more Earth-like, so that people and plants would not be stuck inside pressurized enclosures. Ideas proposed for terraforming Mars—converting its atmosphere so that Earth life-forms could survive on its surface—first require thickening the atmosphere and raising its temperature. This could be started with the introduction of artificial greenhouse gases which would have to be manufactured and released into the air. These gases would raise the temperature of the surface, which would free carbon dioxide locked up in the soil. As the atmosphere thickened with the new CO_2, the heat-trapping properties of the additional gas would help warm the atmosphere even more. The hope is that this would be a self-perpetuating process, where the increased heat would release even more CO_2 into the atmosphere, which would heat the atmosphere even more, and so on. According to planetary scientist Chris McKay, even with a runaway process and given the efficiencies of heat-trapping gases, it would take about 100 years to get the average surface temperature close to the melting point of ice. Once the planet is warm and thick enough for the ice caps to melt and be a source of liquid water, the composition of the atmosphere would have to be re-engineered again. This time, oxygen has to be created for animal life to breathe. This could be done by seeding the surface with specialized bacteria that takes in the carbon dioxide in the air, which when combined with water and sunlight results in biomass (growth of more bacteria), and oxygen exhaled out as a waste gas. Again according to estimates by McKay, this process could take 170,000 years to generate enough oxygen for mammalian life—although one could imagine that advances in genetic engineering could speed up this process by a factor of two.

But is there enough carbon dioxide locked up in the surface soil of Mars to provide the runaway greenhouse effect? A 2018 paper by Bruce Jakosky and Christopher Edwards used the latest spacecraft discoveries to tally the various sources of CO_2 on the Martian surface. This includes CO_2 locked up in the polar caps, bound up inside minerals, and adhering to the surface of dust and pebbles in the soil. They find if you unbind every molecule of CO_2 in the topsoil and release it into the air, the pressure of the atmosphere goes from 0.6 percent to 6.9 percent of Earth's. Because sunlight is only 43 percent as strong on Mars as on Earth, the CO_2 pressure would have to be roughly equal to Earth's atmospheric pressure in order to warm the planet enough for liquid water to exist on its surface. Releasing all of the available CO_2 in the topsoil of Mars would thus result in an atmosphere that is still too weak to warm up the planet by a factor of 14 times.

What about the artificial heat trapping gases suggested earlier for jump-starting the greenhouse warming? One scheme that has been advocated for pre-heating the atmosphere is through the manufacture and release of tons of chlorofluorocarbons and perfluorocarbons, which can have up to 20,000 times the heat-trapping capacity of CO_2. But these chemicals do not last forever in the atmosphere. Although Mars is farther from the Sun, its lack of oxygen and a protective ozone layer means that UV radiation will eventually break apart these big molecules. No one as yet has developed detailed models showing what the expected lifetimes of these chemicals will be in the Martian atmosphere. But even if they are in the thousands of years, factories will have to continue pumping such chemicals out in perpetuity to keep the atmosphere warm enough for life.

There is also the matter of nitrogen which is a necessary ingredient for all life on Earth. As stated earlier, there has to be enough free oxygen that it can be fixed by bacteria into a form that can be ingested by plants. The amount of nitrogen available is still uncertain, although measurements from the Curiosity rover suggest nitrogen compounds are found to be 0.01 to 0.1 percent of the Martian soil by weight. Mars' surface rock would have to be strip-mined and processed to free the available nitrogen into the air (as well as allowing access to other useful bound-up molecules like water). Alternatively, if there is not enough accessible nitrogen on the surface, it would have to be imported in from elsewhere in the solar system, such as Saturn's moon Titan.

Finally, a buffer against radiation from space will not begin to build up until the atmosphere is thick, dense, and oxygenated. If there is enough oxygen, the Sun's UV rays will convert part of it into ozone, where it can shield living organisms on the surface from the DNA-damaging UV radiation. The technology to create a magnetic field in a planet that does not already have one is far beyond what we can imagine today. So for now, radiation shielding on Mars will always be limited compared to Earth, and can only grow slowly as part of a hundred thousand year-long terraforming effort.

Colonizing Earth-Like Worlds

What about the possibility of finding and colonizing worlds elsewhere in the universe that are already Earth-like? Having easy access to worlds with Earth-like surface gravity and breathable air is a primary unspoken premise for *Star Trek*, as well as virtually any science fiction film and TV show where humans traveling out among the stars is a given.

The science is on its way to supporting parts of this premise. Since the first discovery of planets around other ordinary stars in 1995, there have been over 4000 confirmed planets as of this writing. Based on the statistics of the observed stellar sample from the Kepler spacecraft, it is thought that there could be tens of billions of Earth-sized worlds orbiting in the habitable zones of their stars. That is, these would be planets that are orbiting neither too close and too hot, nor too far and too cold for liquid water to exist on their surface. There is no guarantee that every one of these would have running water on their surfaces. Technically, Mars and Venus are in the Sun's habitable zone as well, and they have evolved to be inhospitable places. But assuming we can identify planets that have conditions conducive to life as we know it, how hard would it be to colonize one if we assume, like *Interstellar*, that getting there is no problem?

If we find an Earth-like world, it will either be lifeless or full of life. If the world already has life, the best situation would be one where pho-

Edmunds planet, which in the film *Interstellar* is habitable and promising for human colonization (Paramount/Warner Bros.).

tosynthesizing organisms already exist. They would have filled the atmosphere with oxygen and negate the need for artificial terraforming. First we would have to ignore the ethical quandary of inserting our own alien biology into an existing ecosystem. We would become invading colonizers, with all the baggage that term holds in human history. That means if there were elements of the alien biosphere that were dangerous or incompatible with Earth life, the alien life would have to be cut back or destroyed in order for the invading Earthlings to flourish.

How terrestrial life interacts with the alien biology is a more complicated and important question. This is because humans are not just defined by the cells containing human DNA, but also by the host of symbiotic microorganisms that are part of our "microbiome" that we carry with us. It is estimated that in our bodies, bacterial and viral cells actually outnumber human cells by a factor of 10 to 1. These cells are found throughout the interior of our body, as well as on the surface of our skin. How they interact with human cells as well as the microbiology in our environment is an active area of research which we are only beginning to understand. For instance, there is now evidence that the types of bacteria that we carry with us can determine how susceptible we are to certain diseases, and even how easy it is for us to lose weight. In order to have a healthy human population, it may be necessary to bring along all of these microbiomes to off-Earth colonies. How will such ensembles of Earth bacteria and viruses interact with their alien counterparts? Will they get along? Or will one type of biology overwhelm and wipe out the other one? If the alien bugs have an upper hand over Earth bugs, what does that mean for human health and well-being? What about their impact on the growth of crops and other vegetation? Will we have to come up with vaccines to inoculate ourselves and other Earth-based life against alien bacteria? We won't know until we run such an experiment.

On the other hand, if the world we encounter is lifeless, the space travelers likely will have to run the same terraforming process as described earlier for Mars. The fifth of Earth's atmosphere made up of oxygen molecules is out of equilibrium with the Earth. It is generated by plant-life and if the plants were to disappear, the oxygen would slowly combine with elements and be re-absorbed back into the Earth's surface until there was no free oxygen left. An Earth-like world where all the plant-life was wiped out would slowly turn Mars-like, with the oxygen replaced by carbon dioxide. Newly arrived settlers to a lifeless world would have to start the oxygenation process themselves, by going down the long road of terraforming, and seeding the planet with oxygen-generating bacteria. They would again have to wait thousands of years for oxygen to build up to a level making the air safe to breathe.

One common theme in science fiction is the utopian future where the destiny of humanity is in the stars. This *Star Trek*–like future has us spreading out into the galaxy with ease, exploring where no one has gone before. But there is also a dystopian counter to that ideal, where it is assumed that as Earth becomes trashed and polluted, we have to escape it to survive. There are many examples in popular cinema, including the Off World Colonies advertised in *Blade Runner*, an Earth that is a "shithole" not worth returning to in *Alien: Resurrection*, and an Earth that has collapsed under the weight of rampant consumerism that *Wall-E* is attempting to tidy up.

But as we have seen, colonizing other planets is not easy, and involves enormous scientific and engineering challenges. It can not be done quickly, nor is it likely to be cheap. Planetary scientists who are pro-terraforming understand that with the technologies that we have today, it would still take thousands of years to lead to a viable biosphere where people could live without spacesuits. And even if like Anne Hathaway's *Interstellar* character, we happen to find an Earth-like planet with an oxygen atmosphere, colonizing such a world would involve the unpredictable scenario of mixing together two possibly incompatible microbiological ecosystems to see what will happen. And if not, we face the same terraforming challenges we had seen for Mars.

There is a subtext to space colonization that it is an easy way out if we manage to make Earth unlivable. But it turns out that *Earth life is really suited for planet Earth*. One cannot easily discount the scientific, engineering, financial, and temporal challenges when trying to replicate Earth conditions elsewhere in the universe. There is no Planet B we can easily escape to in case things become unbearable on Earth. In order to generate the resources, technologies, and time that would allow us to terraform and colonize other planets, we have to buy time for us here on Earth.

Notes

1. For more about Earth life surviving on Mars, see my interview with Steve Lee in the final chapter on *The Martian*.

2. However, there is enough variability that the temperature can be as high as 20 deg C (or 68 deg F) at the equator at noon, to -155 deg C (-247 deg F) at the poles.

3. For another take on perchlorates and growing food on Mars, again see the interview with Steve Lee about *The Martian* at the end of the book.

Chapter 3: *2001: A Space Odyssey*

The Sun, Earth, and Moon align in the opening shot of *2001: A Space Odyssey* (MGM, 1969).

Stanley Kubrick is one of the greatest directors in the history of cinema. Born in 1928, as a youth he was mostly interested in reading books, watching the New York Yankees, listening to jazz, and playing chess. His first passion was photography and that eventually led him to the movies. His early films were self-funded, independent, and mostly fictional narrative films with an experimental bent. His big break came when Kirk Douglas asked him to step in and direct *Spartacus* (1960) after Anthony Mann was fired. Kubrick became the youngest director, at 31, to make an epic (it was one of the most expensive movies ever made in the U.S. at the time). The film was a critical and commercial success, and it marked the arrival of Kubrick as a major director. He would go on to direct to direct 8 more films in his life (in 39 years), taking his time and working in a very methodical and meticulous manner with each of his films: *Lolita* (1962), *Dr. Strangelove* (1964), *2001: A Space Odyssey* (1968), *A Clockwork Orange* (1971), *Barry Lyndon* (1975), *The Shining* (1980), *Full Metal Jacket* (1987), and *Eyes Wide Shut* (1999). His films would become landmarks of cinema and an inspiration to several generations of directors. He is truly one of the greats, and along the way, he even helped to inject life into a lackluster industry at its lowest point in history.

The '60s were a trying decade in the history of cinema. After the breakup of the studio system in a 1948 Supreme Court decision called the Paramount Decree, the old Hollywood would be no more. The studios had been practicing something called "vertical integration" for decades: the studios virtually controlled everything: production, distribution, and exhibition. They even "owned" the people: the directors, actors, editors, cinematographers, etc., were all under contract to a specific studio. Directors might finish up a love story on a Friday and begin work on a science fiction film on the following Monday. Actors were famously loaned out to other studios (for a price) or traded for another actor (like so many baseball players). The Supreme Court finally ruled that it was a monopoly and had to be broken up. It took several years, but that original studio system monopoly would finally fall. That was one domino in a larger paradigm shift for a changing nation with a changing demographic.

There would be several other factors that would lead to a decline in movie attendance coupled with film production: the mass migration from the cities (where the theaters were located) to the new suburbs in the wake of World War II, the advent of TV which became direct competition to the movies, the strangulation of movies by the Hays Code (the self-censorship code that had been in effect since 1934), and the influx of foreign films with more artistic content that were not necessarily bound by the Code. Add to that a changing demographic with the baby boomers coming of age and a counterculture which had no interest in saccharine love stories and stagey productions, and the grand result was that movie-going and movie production were at all-time lows. Many thought Hollywood was finished as the '50s turned into the '60s. There were two films that kept Hollywood going through this period, however: Hitchcock's *Psycho* (1960), and Kubrick's *2001: A Space Odyssey*. Both would inject life into a hapless industry.

Psycho brought people back to the movies in new and innovative ways: publishing specific show times (a new practice), not allowing anyone into the show once it began, and showing a trailer where Hitchcock himself urged the audience not to tell anyone else what happens in the movie. Scared and exhilarated audiences would burst through the doors after a show and those standing in the line for the next show would be chomping at the bit for their turn. It became an "event" just to go to the show. Of course, the film itself not only pushed the boundaries of the aging Hays Code, but it stunned audiences with its low-budget horror and rule-breaking narrative. Taking showers would never be the same.

Psycho would pump life into the cinema, but overall attendance lagged throughout the '60s. By the time Kubrick decided to make *2001: A Space Odyssey*, he was under pressure to produce a hit for storied yet ailing MGM Studios. Many thought the giant would go out of business otherwise. Still, Kubrick took his time, spending four years in production culminating with an extremely meticulous shoot. The majority of the film was shot outside London at Shepperton Studios, except for a second-unit journey to Africa for the opening sequence. At a final cost of around $12 million and with a ton of anticipation, the film premiered on April 2, 1968, in Washington, D.C. At first, audiences didn't know that to make of it and it was both a box-office and critical dud. The anxiety over the film, the fate of MGM, and the film industry writ large would hang in the balance. All would have a happy ending.

The film was loosely based on prolific sci-fi author Arthur C. Clarke's short story "The Sentinel." Clarke and Kubrick used the story as a starting point for their script, and they wrote the script together (simultaneously as Clarke was writing the novel) over the course of three years. The central premise of the story concerned a lunar expedition finding a mysterious, alien artifact on the moon; that artifact was sending a powerful signal toward Jupiter once it was found. This would provide the foundation of the film, but as with everything else he did, Kubrick would change it, mold it, and make it his own—and many times, it would be difficult to follow. Part of the initial audience and critical trepidation toward the film had to do with its oblique narrative, part of it had to do with the marketing of the film, and part of it had to do with the changes in film and filmgoing audiences of its time. So it is therefore necessary to start with an outline of the narrative—which may be its least important aspect—before we can sort out all of the historical, cultural, technical, and thematic aspects. Then we can understand the evolution of its reception.

The opening sequence "The Dawn of Man" starts after an ominous symphonic overture, a black screen, and then the Earth rising from the perspective of the Moon. This placement will become more evident, and more consequential, later in the film, after humans find a second monolith on the Moon (the first was found by a group of apes in the beginning of the film). It is as if the aliens were watching the Earth and its first beings in this opening shot, and that it is taken from their perspective. Back on Earth, in prehistoric Africa millions of years ago, we see a tribe of apes struggling to survive, non-violently conflicting with another tribe, and cowering in a cave at night. They wake up to a mysterious black monolith, the leader "Moonwatcher" touches it, and he is seemingly inspired to turn a bone into a weapon, with which he beats a member of the opposing tribe to death. A match cut from the bone to an orbiting nuclear satellite millions of years later signifies a thematic match as well as the object match: both "tools" came to be used as weapons. In one sense, Kubrick tells us that not much has changed across the millennia. The matter of the monolith suggests the alien race (or higher power? Or a future, time-traveling "us"?) placed the monolith as a technological nudge, moving the species forward. One of the most famous cuts in the history of cinema has meaning on multiple levels.

The second, untitled sequence finds bureaucrat Floyd on a shuttle to the moon, where another monolith has been unearthed. We also get the first lines of dialog in the film, about 25 minutes in, as a flight attendant speaks to Floyd. The second monolith is a progress marker: once the civilization is advanced enough to find it, it then emits a signal to Jupiter, to the next monolith. Arthur C. Clarke called it a "burglar alarm" that goes off, literally in an ear-piercing "scream," as the sun touches the uncovered monolith. The civilization has now developed the requisite technology to get to Jupiter, and 18 months later, that mission is in progress, moving us to the third episode of the film: "Jupiter Mission." It is here where we meet Frank and Dave, the astronauts on the mission (others are in stasis), but also HAL, the onboard AI who is helping run the show and who soon turns dangerous. Once they arrive at Jupiter, after the rest of the crew is killed and HAL disabled, Dave sets off on his own to follow the signal emitted by the third monolith. This last episode, "Jupiter and Beyond the Infinite," begins. His journey through a visually spectacular wormhole lands him on/in what seems to be an alien planet (and what looks to be a luxury suite hotel room). As Dave ages in a series of cuts, he dies and is changed (reborn?) into a fetus. A zoom into the last monolith (at the foot of his bed) cuts to the floating fetus orbiting Earth. That's all.

But this rendering of the plot does little more than explain the narrative and is barely the starting point. The *mise-en-scène*, cinematography, editing, and sound, coupled with the rich and thematically dense interpretations of the film, is where the fun begins. Kubrick is such a master at all of these elements, and then some, that there is a lot to sort through; here we go.

Let's start with the *mise-en-scène*. Everything is so meticulous that we can take any shot, scene, or sequence, and fill up the entire chapter. The first two sequences are particularly striking in their authentic atmospherics—we move from the stark beauty of Africa to the stark beauty of near-Earth space in a seamless and thematically consistent flow. The setting of Africa is strikingly prehistoric and simple, but it still highlights the importance of our human lineage, the emotions and actions of early hominids and how that translates to contemporary beings, and it shows how we possess not only compassion and care for our young, but also how we can be violent and unpredictable beings. The sparseness, ruggedness, and simplicity of the setting speaks to everything we were, are, and how we have come to be. Then, the transition to space is connected by more than just a bone to a satellite; it is a cut from a seemingly sparse yet unknown space to another seemingly sparse, unknown space. Neither are entirely sparse, however.

The subjects of each setting are also compared in these opening sequences. The apes of the dusty and rocky landscape huddle together in a cave at night, frightened by the sounds of the animals who may hunt *them*. The humans of the "present" travel alone. The apes guard their young and aim to protect and guide them from the dangers of the leopards stalking the prairie. The humans of the present speak to their young through mediation—video phones or recorded messages—and are seemingly disconnected from them on many levels. The idea here is that we have come so far technologically, but we are also becoming disconnected socially. As technology evolves and we become increasingly reliant on it, we tend to lose our humanity and become more machine-like in our own evolution. Take a second to remember that this film was released in 1968 and consider this comment. That theme of disconnection and losing our humanity in the era of technology and the age of machines would pervade the film and find its voice in all of the formalist aspects. And it is spectacularly prescient.

The cinematography, however, may be (justifiably) the most remarkable aspect of the film. Geoffrey Unsworth, one of the more prolific and celebrated cinematographers of all time (*A Night to Remember* [1958], *Cabaret* [1972]) worked tirelessly to realize Kubrick's vision. Like most aspects of every one of his films, Kubrick worked closely with Unsworth on all of the details, especially the visual effects. The film's lone Oscar would be in the category of Best Effects, Special Visual Effects, given to Kubrick. The film's visual poetry is really a marriage between these two aspects, and the painstaking approach to each is obvious upon viewing the film.

The opening sequence in Africa, while it may seem simple, was actually a complex *mélange* of camera, live action, and special effects, and it serves as a microcosm for the film. It starts with a second-unit shoot in Namibia who filmed a host of different landscapes (with Kubrick directing them from the phone since he hated to fly!). The next phase took place in the studio, where those shots were projected onto a screen behind the actors (in ape costumes). The challenges here are obvious—how to blend both the foreground and background so the sequence looked as real as possible. Kubrick decided to eschew the common practice of the period—using rear projection for the background—so he could attain a more realistic sequence. Using front-projection had its obvious technical issues: how to achieve the visual clarity when the projectors are so far from the screen in the background and then shot by the camera for the actual film sequence; and, perhaps even more difficult, how to account for the shadows of the actors and animals. This is where cinematography and visual effects come together, and this is where Kubrick shined. He loved the challenge.

The first issue was how to get a clear picture on the projected background. The solution for this was two-fold: a strong projected image as well as a screen that could capture that strong projected image. As with many challenges on the film, they used what they had and invented what they didn't. They wound up projecting the stills from an 8 × 10 plate onto an entire wall, 40 feet × 90 feet, something that had never been done before. They had to keep fans running constantly due to the intense heat needed to generate the light from the projector. Then they invented a reflective screen that could hold the image while being photographed by the actual camera. They used mirrors to angle the projection onto the screen with the camera filming behind the mirrors and adjusted it so perfectly that there were no shadows from the actors nor animals. The one shot where we get a sense for lights and camera is when the leopard looks toward the camera and her eyes shine; it is an eerie and ominous shot that Kubrick left in, despite its break from realism. While this may be a short and sweet précis of the entire process, and it is much more complex than this simple description, it serves to show how complicated it all was but also innovative and proficient a crew they were. They crew went through a journey perhaps as difficult as the astronauts': as legendary special effects master Douglas Trumbull (*Close Encounters of the Third Kind, Star Trek: The Movie, Blade Runner*) noted, he signed on for a year and finished up three years later! All their hard work shows right from the start of the film.

The result of this opening section is a fascinating and realistic sequence that captures the early hominids in their environment while infusing the film with a sense of history, human connection, and technology from the very beginning. For all of the film's thematic insistence on disconnection and the lack of humanity, it also has a great deal of heart. Humans have humanity, even in their earliest forms. The next sequence—in space, after the match cut from bone to satellite—also combines the beautiful mix of old and new technologies that we see in that first sequence, yet we start to get a very different view of humanity.

Models had been the foundational cornerstone of special effects since the nascent era of film, going back to George Méliès and finding its footing in the brilliant and groundbreaking special effects work in *King Kong*. The models are used for all of the spaceship exteriors, and it is here that cinematography and special effects meet. Intricate camera movements, combined with complex lighting illuminating the ships and the stars, allows for a striking sense of realism in the pre–Moon landing era (the moon landing was still two years away as the crew worked in London). Kubrick constantly strove for realism in the space sequences and it shows from the start as we see Floyd taking a shuttle to the moon.

The trip to the space station, cut to the glorious soundtrack of "Blue Danube," is pure visual poetry. From the moment we hear the music, it conjures up an ageless feeling of wonder, custom, grandeur, and a sense of normality/continuity. Yet Floyd, asleep on the shuttle, is immune to the flight and the beauty outside the window. He is commuting to work, as if this is normal and just another day at the office. The music, seemingly contrapuntal, adds depth and heft to the sequence while giving it a feeling of timeless beauty. The space station comes into view and is a wonderfully spinning visual marvel in a geosynchronous orbit, a sight to behold on the outside as well as in. This first view of an actual future-world of space populated with humans is stirring and moving. But the docking sequence—long, slow, and methodical—is a dance in and of itself. The phallic-shaped shuttle approaches the opening of the landing port as if engaging in a foreplay among machines, consummated only after just enough dancing and parrying. It is not the first time the machines would become anthropomorphized, enjoying themselves and seeming to have much more fun than the human beings. When Floyd arrives on the station and calls his daughter on the vid-phone (imagine! a phone call where we can see the person!), she tells him that she wants a telephone for her birthday. The banal call and discussion presages every conversation between child and parent of the actual 2001-era and beyond: a request for a phone. Any parent of millennials has been there.

The space station itself has a stark yet stylish minimalism to it on the interior—here the *mise-en-scène* gives us a sleek, futuristic lounge populated by both Americans and Russians (as the Russians pump Floyd for information about his visit). Kubrick shows his prophetic touch here, in both the historical and economic contexts, envisioning a world where the Cold War has ended and space is traveled by many. The Station is home to a Howard Johnson's and a Hilton showing the win for capitalism in the future. One great story about the film shoot in London involves Russian diplomats visiting the Shepperton Studios. As Kubrick gave them a tour of the model spaceship, the Russians were impressed—with one exception. "This looks very authentic," said one of the visitors, "except in the future, the instructions will obviously be in Russian, not English." Perhaps he was wrong.

Floyd then takes a smaller shuttle to the moon (again, sleeping through most of it), where we get a great, extended sequence of the flight attendant seemingly walking upside down to enter the cockpit. The shot was accomplished by the camera rotating 180 degrees, rather than moving the set or any other trickery. It is a wonderfully seamless shot where it is impossible to tell there is any camera movement. On the shuttle, the crew eat from prepared food boxes with straws sticking out (to allow for the weightless environment), and the flight attendants walk with Velcro shoes in the zero-G. It is another long sequence with long takes that allow us to breathe in the film and the atmosphere, heightening the realism of the voyage.

That brings us to the editing, which moves from wonderfully reflexive transitions such as the bone/satellite match cut to long takes that immerse us in the shot and allow us to experience the journey right along with the subjects. The cinematic move here is from formalism to realism; the match cut is a formalist technique—it announces to us that it is a film, a construction, and in this case it accounts for Kubrick at the helm. It is infused with meaning that Kubrick asks us to consider. The long takes—shots that last for a long period of time without editing—don't announce anything or ask us to consider anything; they allow us to sit and consider the experience as is. It is a realist technique. The long take of the machines copulating or the flight attendant doing her job in space allow us to soak in the journey. The formalism and the realism live together happily in this film and both have a purpose. Kubrick masterfully weaves both editing techniques into the film, asking us to intermittently engage on an intellectual level while also allowing us the space to enjoy the moment. It is Kubrick's superpower.

The sound of the film, including the score as well as the extended sequences of silence, is also justly famous. The use of classical music was a choice Kubrick made late in the process. He thought the classical pieces would make it more timeless. He even commissioned a whole score from composer Alex North before he made a final decision, ultimately choosing not to use it. (A side note here: North didn't know Kubrick abandoned it and went to the premiere unaware. He walked out in disgust as soon as he heard he classical music score.) But you can't argue with the results. "It's hard to imagine anything other than the wonderful classic score," actor Gary Lockwood (Frank) notes: "It gives the film a timeless feel and ensures it will never go out of style." The visuals and the music are in sync in a way hardly ever matched in the history of cinema. The opening of the film is set to Richard Strauss' "Also Sprach Zarathustra," and with its allusions to Nietzsche's *Thus Spoke Zarathustra*, the allegory and mythmaking begin from the first note. As Arthur C. Clarke has said, in a very succinct and concise manner that Kubrick would never condone when labeling the film, "it is a story about evolution." Both Strauss and Nietzsche had this in mind with their respective works. More on Nietzsche later, but the opening from Strauss, covering the images of the Sun, Earth, and Moon in alignment, give us the impression of beginning times on the planet (without the primordial ooze). We also get other famous pieces throughout the film, including "Blue Danube" and Khachaturian's "Gayane Ballet Suite," as well as more modernist compositions such as Ligeti's "Requiem," "Atmospheres," and "Lux Aeterna." Again, it is a mix of the classical and the modern, just as the cinematics intertwine formalism with realism. Aside from the score, the other sound in the film is equally masterful: the long sequences lacking any dialogue, the negation of sound in space, and the emotionless voices of the astronauts as well as the mellifluous voice of the most famous character in the film, HAL. HAL is infused with more humanity than the humans, and even while killing off the sleeping crew and unsuspecting Frank, he seems more alive than the living beings. Each aspect of sound—from the simple to the (at times) wallpapered score—is effective and even rhetorical. In the final analysis of the cinematic aspects, the four constituent elements of film, the *mise-en-scène*, the cinematography, the editing, and the sound, are all varied and purposeful and point toward a mix of the classical and the modern, the formalist and the realist, while highlighting the lack of humanity in an age of technology. That brings us to themes.

One of the main subjects of the film is technology vs. humanity. Kubrick seems to be conflicted about this: on the one hand, technology takes humanity to great heights and inspires them to do great things; on the other hand, as the match cut illuminates, we also misuse technology for impertinent and selfish ends, such as weapons. One of the more prominent themes I see in this dense and sprawling film is that *technology de-humanizes people and turns us into machines*. The technology in the film is more human than human, and the advances come from outside humanity. Here are some specific examples of how that looks:

- The first tools, implying scientific thought, are instilled by an outside force, not a real human accomplishment. We are "nudged" by Others.
- Astronauts and scientists are emotionless and engage in joyless work without much interaction. How many times do we see Frank or Dave smile? Even the conversations with family are puerile and antiseptic.
- The food is terrible—it is a world bereft of sensation or pleasure. At one point, Floyd points out that "They are getting better at this" as he eats a sandwich on the lunar rover. And when we consider our contemporary society, few people cook, and many eat the same,

bland, chemically altered food. Kubrick was right about the many things in his assessment of what the year 2001 would look like.
- HAL has more emotion than the astronauts, and he even tries to coax emotion and conversation out of the listless humans, serving as psychologist and friend during the mission.
- Even Frank's death is met with a cold, detached, and business-like reaction and countenance from Dave. He just goes about his business as he tries to save Frank and then when faced with the futility of the task, releases him into space without a peep.
- The spaceships even "copulate" in the extended docking sequence at the orbiting space station. Humans never come close to such affection much less eroticism.
- The comment on the lack of human interaction in the world is never more evident than when the little girl asks her father for a telephone for her birthday. (Again, how right on was Kubrick about the future!)

Overall, the message seems to be that technology dilutes our humanity, our ability to connect with each other, and our ability to communicate on any type of emotional level. And if we point back to our central question of science fiction: "What does it mean to be human?," we find our answer layered throughout the film. One layer points directly back to Nietzsche.

In Nietzsche's *Thus Spoke Zarathustra*, one of the ideas he posited was that man had to overcome God before elevating to the level of superman, or *übermensch* (a term that is now unfortunately fraught with sexist, nationalist, and racist overtones). The idea, however, was that this next level of humanity would be enlightened and find the true meaning of life after a progression from animal to man to superman. Of course there is much more to Nietzsche's work and this is a gross simplification, but for our purposes, it should suffice. In terms of the film, we see the transition from animal (Moonwatcher) to man (Floyd) to superman (Bowman) that includes Bowman overcoming "God" as he kills HAL. We can interpret the star-child at the end of the film, then, as that move to a superman and enlightenment. The point is that we should fight technological dehumanization, and what it means to be human, simply, is to keep our connectedness and our humanity. Simply.

The film was not well-received upon its initial theater run, and it was on the verge of being pulled—with great consequences for everyone involved. The reviews were mixed, and the film was alternately called "oblique," "confusing," and even "pretentious." But audiences had something else in mind, and they kept showing up and showing up. The numbers multiplied quickly, and theater owners recounted tales of young, drug-induced hordes sitting on the floor in front of the screen and getting lost in "the trip." Theater posters changed quickly, showing the star-child with a tag line of "The Ultimate Trip." The film wound up staying in theaters and ultimately became a financial success. It was nominated for four Academy Awards, including Best Director, Best Original Screenplay, and Best Production Design, winning for Best Visual Effects. Carol Reed won the Best Director Award for *Oliver!* (1968), and *2001: A Space Odyssey* wasn't even nominated for Best Picture. *Oliver!* won that as well.

The mid-'60s in the realm of science fiction cinema was somewhat akin to the search for a white whale. Directors were Ahab, albeit not on a revenge tour, but rather yearning to do what no one else could/had done: they were all looking to make the great science fiction film. Even though there were a few examples of "adult" sci-films in the '50s, including *The Day the Earth Stood Still* and *Invasion of the Body Snatchers*, most sci-fi never made it above the B-realm and remained mired in tin robots and oversimplified bad vs. good scenarios. The new generation of post-classic Hollywood cinema wished to change that dynamic, especially considering the dearth of serious sci-fi literature in the post–World War II era. Since American cinema was at a low point and the European cinema so strong, much of this sailing took place in Europe, particularly in the "art cinema" scene. Jean-Luc Godard made *Alphaville*, and François Truffaut *Fahrenheit 451*. And remember that the film-in-the-film in Fellini's *8½* (1962) was a science fiction film where a select group of people blast off from Earth after an apocalypse. That the premise was so ludicrous and potentially expensive was not lost on Fellini. Yet Fellini said that after seeing *2001: A Spacey Odyssey*, he was "cured" and no longer needed to chase the whale. John Lennon, after seeing the film for the first time, remarked: "It should be played in a temple twenty-four hours a day." Science fiction had grown up.

—Vincent Piturro

Catching HAL

Ka Chun Yu

Director Stanley Kubrick's and science fiction author Arthur C. Clarke's collaboration on *2001: A Space Odyssey* created a sensational depiction of what space travel would be like in the near future. The film was released in 1968, a year before humans first landed on the Moon. It extrapolated from the space race between the United States and the Soviet Union to show traveling into space to be as commonplace as boarding an airplane flight, gigantic spinning space stations, bases on the Moon, and a manned mission to Jupiter. However, today, none of these space-based technological developments has come to pass, even two decades after the year 2001.

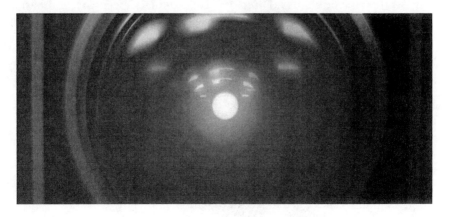

HAL 9000 from *2001: A Space Odyssey* (MGM).

Yet *2001* also introduced the HAL 9000 computer into our pop culture consciousness. HAL was an artificial intelligence (A.I.) that felt so human that it is impossible to imagine HAL while reading Clarke's novel today without hearing Douglas Rain's voice. HAL spoke in conversational language with the human crew, had interactions that were far from machine-like, and as the end came, solicited sympathy from the audience as astronaut Dave Bowman disconnected his circuits despite his pleas. (And because HAL has so much personality and passes the Turing test—when a machine in its verbal interactions is indistinguishable from a human—I will refer to HAL as "he" in this chapter.)

HAL has many similarities to the A.I.s that are becoming more important in our everyday lives. Are we close to creating A.I. similar to what is seen in that film? Is the cinematic HAL merely a slightly more advanced version of the real-life Watson that competed on the TV game show *Jeopardy*, or Siri that listens for our requests on our iPhones, or Alexa that does the same in our homes? The answer is that HAL is far more advanced than any of those real-world technologies. HAL thinks, schemes, and carries out plans more like a human being than any computer program that we have today. Moreover, the A.I. services being developed today by Google, Facebook, and Microsoft are built on machine learning which is more restrictive than the type of learning done by humans. Although highly successful in the many realms where they operate, our current day A.I.s are far from perfect. So like moonbases and spacecraft with centrifugal living spaces,

HAL-like A.I.s are another example of a real-world technology that does not match its fictional counterpart. To see why, we must first learn how A.I. in the real world works, and then we will see what it will take to catch up to HAL in *2001*.

In the film, we see HAL beating a human at chess, something that computers in our world have no problems doing. But HAL also demonstrates skills that seemed like magic until recently. These are activities that computer scientists are continuing to advance or develop for today's real-world A.I.s. They include facial recognition to distinguish different human beings, speech recognition to identify spoken words, natural language processing to understand the meaning of that speech, and expressive speech to communicate back to the astronauts. All of these capabilities are ones that A.I.s today can handle with varying levels of success in our world.

How does real-world A.I. work? Let's look at one type of skill that A.I.s can perform pretty well today: image identification. If you perform a Google search for pictures of cats, you will be served an endless number of feline images that have been posted by their owners.[1] There are no human beings involved in tracking and tagging all of the cat pictures that Google keeps track of. Instead, machine learning algorithms have been trained to identify cats (and countless other objects) by being fed thousands of pictures of cats, and each time being told it was viewing a picture of a cat.

This type of machine learning requires software to simulate an artificial network of nodes for identifying patterns, similar to how a network of neurons in the human (or animal) brain might work. (Although these "nodes" are similar to nerve cells because they connect to and communicate with each other, the real brain and the nerve cells that make it up do not operate exactly the same way.) This "neural network" communicates with the outside world via an input layer of nodes (the far-left column of nodes in the below diagram) and an output layer (the far-right column of nodes). There are also multiple hidden layers in between the input and output layers. Actual neural networks in use will have different numbers of nodes at each layer and will not be restricted to just four hidden layers. However, all neural networks will have an input and output layer. The specific details of how a neural network is set up and operates will be different from one task to the next, but the description given here is a good generalization of the process.

As shown by the connecting arrows, each node in a layer is affected by the nodes from the previous layer. Each arrow has an associated value that represents the strength (or the "weight") of the connection between each node, with values ranging from a minimum 0 (fully "off") to a maximum 1 (fully "on"), and every possible value in between. A weight of 0 will have no effect on the node downstream, while a weight of 1 will have a full effect.

When a picture of a cat is fed into the neural network, the pixels representing that picture are decomposed into signals that then feed into the nodes in the input layer. Depending on the weights of the connecting arrows, the values of the signals change as they cascade from left to right, with the end result popping out in one of two nodes in the output layer.

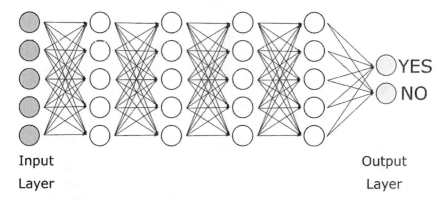

A neural network is set up with the input layer on the left, and an output layer on the right. Information from the image is sent between different nodes in the different layers. "Weights" associated with each arrow modifies each input as it moves along from one node to the next, from left to right. At the end, the output nodes for this neural network can be used to answer a question with two answers, such as "yes" or "no" (Ka Chun Yu).

A neural network is set up with the input layer on the left, and an output layer on the right. Information from the image is sent between different nodes in the different layers. "Weights" associated with each arrow modifies each input as it moves along from one node to the next, from left to right. At the end, the output nodes for this neural network can be used to answer a question with two answers, such as "yes" or "no."

When the neural network is first set up, the weights of the connecting arrows are arbitrary. The chances that a random image of a cat fed in at the left will result in a YES being churned out on the right is 50 percent, because initially, each choice is equally likely. This is where supervised training of the neural network comes in. After an image is fed through, information about the result—whether it was correct or wrong as determined by a human—is sent back into the network using a mathematical sleight-of-hand called "backpropagation." The connecting weights between the nodes adjust as the error correction moves backward from the right to the left. After a backpropagation, the cycle can repeat itself: another picture is sent through from the left to the right, the resultant output is compared to the correct answer as called by a human, and the error correction is backpropagated to the left. As you continue to train the network—typically with thousands if not tens of thousands of pictures—the weights will modify so that the network gradually becomes better at recognizing cats.

A neural net can be used not only to identify cats in general, but to identify individual cats (or for that matter, individual people). Instead of two output nodes signifying YES and NO, you could have five nodes associated with five different cats that you want to single out. Again, you will need to train with many pictures of each cat, and backpropagate the errors into the network for each picture. After the training, you can send a picture of a cat into the input layer that was not in the training set to see how good the neural network is now at identifying the different felines.

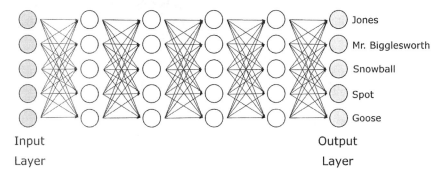

Input Layer

Output Layer

A neural net with five output nodes, which can be trained to recognize five different cats (Ka Chun Yu).

In this way, "deep learning" neural networks (involving multiple hidden layers in between the input and output layers) can be used to identify not only cats, but the countless other objects in our lives that might be the subject of an image search. However, this is just the tip of what machine learning can do. Neural networks are not restricted to making matches in pictures or in video. They are used to look for matches in a wide variety of pattern recognition tasks: detecting letters in cursive handwriting, forecasting what a shopper would purchase based on past shopping behavior, determining how likely that a borrower will default on a loan, or predicting heart attacks from electrocardiogram signals.

Neural networks are extremely powerful tools, but they also have limitations. The only way to improve a neural net is to continue feeding test data vetted by a human. Thus, training a network can be highly time-consuming, since huge numbers of images along with their identifications are needed.[2] When errors are backpropagated, they change the connecting weights between nodes in minute and inscrutable ways. In traditional computer programming, if there is an error in how a software program is working, a programmer can tweak the instructions in the program to improve the code. However, in a neural network, there is no *a priori* way for a human to guess what the connecting weights should be in the network for it to distinguish a cat versus a dog, or any of the other decisions neural networks are asked to make. The only way to improve the neural net is to feed it accurately tagged training data.

Here it becomes obvious that there is a big difference in how neural nets and humans learn. The former require thousands of training examples to lead to improvements in learning. Humans are different. Even a young child can learn to recognize a cat based on a handful of examples. After an infinitesimal fraction of the training that is required of an artificial neural network, that child will subsequently be able to point out cats in real life, in photographs, in *Tom and Jerry* cartoons, in *Hello Kitty* keychains, and as plush dolls in a toy store aisle. Read a picture book to children and afterward, they will identify not only cats, but could point out dogs, gorillas, sheep, cows, rhinos, ostriches, and humpback whales in scenarios outside of the book. By comparison, even after time-sensitive training with thousands of photographs of real cats, a neural network will stumble if it receives a picture of Garfield, the comic strip character. Even if you train your neural network on the pen and ink version of Garfield, it's no guarantee that it will recognize a Garfield toy suctioned to the rear window of your car unless it was trained specifically for that as well.

A neural network does not identify cats because it "knows" what a cat is, in the way that humans know and understand what cats are in real life. That is, we understand not only what cats look like, but also how they appear from different view angles. We understand what makes a cat a cat, even when its physical properties change. We can easily categorize cats as cats even when confronted with examples with different colored coats, with long or no hair, with tails of different lengths, and with body types ranging from skeletal to corpulent. Even if they are missing an ear or a leg, with enough other information (including behavioral—is the cat acting as it should when the sound of a can opener or the light of a laser pointer appears?), our brain's categorization circuits would still put them into the category of "cats."

In contrast, the neural net only "knows" what a cat is based on the collective behavior of the nodes in the network after a picture has been fed in. If the connecting weights ultimately lead to triggering the "Yes it's a cat" output node, then the A.I. recognizes the cat. However, this has occurred only because the features in the picture has led to this conclusion. There is no reasoning about how cat-like the object is, based on its collective physical and behavioral characteristics. (Later on, we will see what type of A.I. is necessary that can handle the collective properties that define a category of objects like *cat*.)

Because the neural network is focused solely on image processing and not on any other characteristics, the limited rules it uses to make a decision cause problems. It turns out that there are many different ways to send image data into the neural network that will trigger the "Yes it's a cat" output, even when they do not represent a picture of anything that humans will recognize as a cat. In fact, this input could appear as random speckles of noise to a biological observer but will fool the silicon-based neural net into misidentifying the subject. In a 2015 research paper by computer scientists Ian Goodfellow, Jonathon Shlens, and Christian Szegedy, a neural network was fooled into identifying a picture of a panda as a gibbon (with over 99 percent confidence!) by just this trick.

Misrepresenting animals is not a big deal in most people's day-to-day lives. However, A.I. is used in pattern matching in scenarios that are deadly serious. For instance, autonomous self-driving cars need to be able to identify street signs in order for them to navigate roads successfully. But as researchers have shown, it is possible to create "graffiti" stickers which when applied to a stop sign will cause a neural network to misidentify it as a "45 MPH" speed limit sign. This type of mischief would lead to chaos on the roads. In addition, image recognition of animals and road signs are not the only realms where malicious attackers could create adversarial data that can fool neural networks. Researchers have discovered ways to trick A.I.s that recognize faces, 3D objects, and speech. Even A.I.s trained to play video games by running millions of simulated matches are susceptible to the injection of adversarial data that seem harmless to a human but can cause the neural net to go awry.

Obviously, HAL in *2001* has none of these problems. From its behavior in the film, HAL acts far more human-like than any A.I. in current existence. For instance, it can distinguish between Frank Poole and Dave Bowman, two of the astronauts onboard the *Discovery* spaceship. HAL has natural conversations with them, and even recognizes Bowman's drawings of the astronauts who are in deep sleep during the journey. HAL is still more human-like, when he conveys interest in the artwork by asking Bowman to hold them closer to his all-seeing eye. We get the sense that HAL was not trained with hours of video feed of each individual astronaut in order to recognize them, nor did Bowman have to draw hundreds of sketches of his fellow shipmates as training data for HAL.

Where HAL really shines compared to current day A.I. is his ability to understand events, and his ability to make and execute plans that were not in his original programming. Once HAL goes mad and decides that Poole and Bowman are plotting to unplug him, HAL begins his own scheming. He decides that the only way to save the mission, which he views as his priority, is to get rid of the human astronauts. He concocts a plan to get Poole

66 The Science of Sci-Fi Cinema

This stop sign, modified by small black and white stickers, would still read as a stop sign by a human driver. An A.I. pattern matching algorithm using neural networks may instead identify this as a speed limit "45 mph" sign (Ivan Evtimov).

out into space with another manufactured failure of the AE-35 electronics unit, and then takes control of his space pod, which he uses to kill Poole. When Bowman flies out in his own pod in an attempt to recover Poole's body, HAL refuses to let Bowman back in. When Bowman explains that he will attempt entry via the airlock, HAL is aware that Bowman had left his helmet back in the bay, and thus, declares that Bowman would not be able to survive the vacuum of space to get through the airlock.

In the movie, HAL not only understands his mission, but he is also highly creative when he comes up with his own plan to get rid of the meddlesome astronauts. (Surely, ramming an astronaut at full speed with the pod and disconnecting the life support from the crew in deep sleep were not part of his original programming!) HAL also has the common sense to know that humans need air to breathe, that there is no air outside the spacecraft, and that if Bowman left his helmet back in the bay, then it would be impossible for him to enter the *Discovery* safely through the airlock. HAL has enough of an understanding of how the spaceship operates that he knows he can keep Bowman out, by ignoring his command to "Open the pod bay doors, HAL."

The types of reasoning, creative planning, and initiative that HAL shows in the film are currently impossible for A.I.s in our world. The most advanced real-world A.I.s can identify multiple objects in a picture. Researchers are creating A.I.s that can identify what actions people are performing in videos, from throwing a shotput to house-cleaning and vacuuming a rug. But these neural networks have no information about how the world operates. As an example of this, artificial intelligence scientist Rodney Brooks compares how a person and a current-day A.I. might interpret a picture of people playing Frisbee in the park. An A.I. neural network can be trained to recognize pictures of people playing Frisbee. It may even be able to identify individual people and determine their number in the image. But what the A.I. does not know about Frisbee playing compared to a human is far greater than the few things the neural net can identify in the picture. The A.I. cannot tell you what shape the Frisbee is, how far a person can throw a Frisbee, whether the Frisbee is edible, roughly how many people can play Frisbee at once, if a 3-month-old can play Frisbee, and whether today's weather is good for playing Frisbee. The A.I. also does not have any information about what type of activity Frisbee playing is— is it a job, a leisure activity, or a punishment? Nor does it have any information about the space it takes place in (where are parks located? do you pay money to play? how do people get to the park?). These and many more questions involve information that requires common sense about everyday life that is unavailable to current machine learning algorithms. Artificial intelligence that is truly human-like—what researchers call "general artificial intelligence"—is still many years in the future in our world.

There are attempts to try to imbue A.I. with common sense. Computer scientist Douglas Lenat has spent the last 35 years working with a team of computer scientists and philosophers to build Cyc (a name derived from the truncation of "encyclopedia"), a vast knowledge database that covers factual assertions as well as the unspoken rules governing those facts. For instance, "Abraham Lincoln was the 16th President of the United States" is a fact coded into Cyc's knowledge base. Information about well-known fictional creations are also encoded, such as, "Frankenstein is a monster," "Dr. Frankenstein is a medical doctor," "Dr. Frankenstein is a male human," and

"Dr. Frankenstein is a German person." Many millions of other facts like this exist about the common, everyday knowledge that is widely shared and understood by people. These facts have to be identified, coded via a special programming language, and then entered into Cyc. Many more are generated daily, so in a sense, building Cyc is a never-ending task.

To make the A.I. useful, Cyc also has statements that are not typically written down, but it is general commonsense knowledge known to everyone. They could include: "The President of the United States of America is a human being," "Human beings are mammals," "Mammals require sleep," and "Most sleeping is done at night." With those statements symbolically expressed, Cyc can logically infer that "Abraham Lincoln slept at night."

These symbolic statements about how the world works seem banal because they *are* banal. They describe ordinary, everyday knowledge that seems so obvious that they are not worth writing down. A Wikipedia entry for Abraham Lincoln would never bother to state that he was a human being and a mammal. However, it is precisely such statements that computers have no awareness of, and which they need to be told about explicitly.

Neural network–based A.I. also has no knowledge about commonsense rules. As an example, Google uses neural networks to interpret search queries in order to determine what results to return. If I type "Who won the 2020 Super Bowl?," Google searches through its saved text copies of millions of webpages. Based on a statistical analysis of those pages that contain the same or similar words as the search query, Google's algorithms will return a list of webpage links that will have a high confidence of answering the question. The person performing the search gets the answer ("Kansas City Chiefs") that they are looking for.

If you enter a question or phrase that has not appeared on a webpage before, then Google has nothing to work with. If you type "Did Abraham Lincoln know how to snowboard?" into Google search, the links returned have nothing to do with Lincoln shredding down the mountain or strapping his boots into a Burton board. Instead, Google returns a random mix of websites that happen to include the words "Lincoln," "snowboard," "know," and "how" somewhere on that page. Google has not found any webpages explicitly answering the question of whether Lincoln ever hit the moguls, and as a result, the search returns are meaningless. Because the search algorithms do not have the common sense to know that snowboarding and Abraham Lincoln do not overlap in time, it cannot provide a sensible answer to the search.

Cyc was commercialized in 2016 as a product of the company Cycorp. Customers contract to create knowledge bases of—among other things—all known terrorist groups and their members, and pharmaceutical names and information from all around the world. Clients collaborate with Cycorp to develop the specialized common-sense rules specific to their knowledge domain. From these rules and the knowledge bases, Cyc will be able to identify possible connections between individual terrorists and terror groups, or how a drug that is commonly known by one name in one country is identified as something else on the other side of the world.

Lenat, the inventor of Cyc, hoped that with a sufficiently large knowledge base, and a wide-ranging enough set of rules, Cyc could start learning by itself. However, to date, there is no evidence yet that Cyc has turned into a general-purpose A.I. that learns like a human without further supervision. For now, the rules-based approach of Cyc is one promising technology that could lead to a future HAL, but we do not appear to be there yet.

One final promising direction for developing general artificial intelligence uses as inspiration the most successful learners that we know about: humans. Or more specifically, human children, who even as babies, are extremely effective at learning about the world around them. Unlike the A.I. technologies that computer scientists have been grappling with, babies are very different learners. As mentioned earlier, neural networks need to be trained on many thousands of images, while young children can get by with far fewer examples. Those examples can also be inexact and messy, and nothing like the curated and processed images that are fed into neural networks. Although parents may play a role in guiding their children's learning, babies often explore by themselves while absorbing information about the environment around them with little to no guidance at all.

Researchers Linda Smith and Michael Gasser think that the reason why infants are so good at learning is that they learn in a *multimodal* way, by using information from multiple senses. For example, newborns have trouble focusing their eyes, but they will still turn their heads towards new sounds, showing that they combine visual and audio information to explore their environment. By three to four months of age, babies will start reaching for objects. They now incorporate the sense of touch into their explorations, but at the same time, they are also noticing how an object changes visually in appearance as it is turned in their hands, and the sounds it makes as it is shaken, thrown, or dropped. As the infant becomes mobile, first by crawling and then by walking, the range of experiences increase even more as she moves past, over, around, and into objects. During these periods of exploration, babies become familiar with their surroundings and the objects that occupy their world. They begin to understand how space is organized, and how objects move inside that space. They learn rudimentary physics, by seeing how objects affect each other as they collide, or are pushed or pulled. They begin to understand cause and effect. With these explorations, young children start to build up the set of commonsense rules that are intuitive to and guide the lives of human adults.

In contrast, a neural network learning to recognize dog pictures is in a far more restrictive learning environment. It is fed only one type of information—pictures of dogs. Because its training uses only the pixel information within those images, the neural network's understanding does not include a dog's three-dimensional form, how it looks from different directions, and its size in the real world. A child growing up with the family dog will become familiar with the tactile sensation of the pet's fur (or claws or teeth) on his skin. He will have first-hand experience of what the weight of the animal (or parts of the animal) means when it leans, sits, or pushes back. The child will build up direct experience of how the dog moves and reacts within the environment, how it smells, and the countless other small details that define dogs from everything else that is non-canine. It is through these experiences that even a young child will have a much more robust understanding of how dogs are different from cats than a neural network trained on thousands of pictures of those two animals.

There are clues in *2001* that HAL is intelligent because he has gone through a prolonged human-like learning process. Near the end of the film, Bowman is disconnecting HAL's circuits, which causes HAL to revert to an earlier, child-like state. HAL says at one point:

> I am a HAL 9000 computer. I became operational at the H.A.L. plant in Urbana, Illinois on the 12th of January 1992. My instructor was Mr. Langley and he taught me to sing a song. If you'd like to hear it, I can sing it for you.

We learn from this bit of dialogue that HAL is nine years old by the time of the events of the film. Today we would never think of buying and using a nine-year-old computer. Because of Moore's Law, computers get faster and cheaper with each passing year. In the film, it is implied that HAL had to take that long to learn about the world. He even had a human tutor who taught him to sing a song, just like a young child who learns from an adult. So HAL is not programmed the same way that we program normal computers today. He also does not learn the same way that neural networks learn by ingesting thousands of picture examples in a short time or running through millions of simulations to learn how to beat human players at a video game.

If HAL learned like a human child, what would that have been like? Before he was uploaded to the *Discovery* spacecraft, HAL would have to be located inside a robot body that could freely explore the world through multiple senses as a human child would. HAL would still have been fed facts and information from the world's encyclopedias and databases (including how to run and operate a complex spacecraft like the *Discovery*). However, much of his understanding of how the world works and the commonsense rules that guide human lives would be learned the same way a human child would learn them. It would be done through personal experience, exploration, and trial and error, while interacting with human tutors.

Computer scientists Gary Marcus and Ernest Davis argue that training A.I.s in ways similar to how we humans learn would have other benefits. Neural networks and the A.I.s portrayed in fiction like *The Matrix* and *Terminator* films do not have a sense of human ethics and morality. If asked to solve the problem of climate change, an A.I. using only logic and statistical analysis might come up with a genocidal plan to exterminate the human race as the optimal solution. However, an A.I. that has an understanding of causality and how the world works could be programmed with rule-based ethical behavior. Science fiction author Isaac Asimov imagined this in his robot stories, where the actions of robots are constrained by the Laws of Robotics. A robot would not be able to make any sense of the First Law of Robots—"A robot may not injure a human being or, through inaction, allow a human being to come to harm"—if it did not have a deep understanding of causality and the consequences of its actions. A robot expected to interact with humans would be dangerous to society if it did not grasp the commonsense rules that guide human behavior.

If HAL was initially raised like a human child, one could imagine that being uploaded into *Discovery* could have been a terrifying and traumatic experience. He would have gone from an existence where he moved freely to explore the world rich in stimuli around him in his robot body, while interacting and learning from his human instructors. After being transferred into the Discovery's computer core, he may gain new sensory information about the operation of the spacecraft, but except for vision and hearing, HAL would have all of his other human-like senses disrupted or disabled. Instead of having the freedom to move and explore, his intellect would be imprisoned inside a sterile spaceship, floating in the blackness of space.

Perhaps there were human psychologists specializing in A.I. therapy who helped HAL transition to his new life. Perhaps the 9000 computers were programmed to feel a sense of pride and accomplishment when graduating to new jobs, even if they restricted their mobility and dulled their perceptions. When asked by the reporter if his responsibilities of running the ship and keeping the human crew alive makes him doubt his confidence, HAL replies with a hint of pride:

> The 9000 series is the most reliable computer ever made. No 9000 computer has ever made a mistake or distorted information. We are all, by any practical definition of the words, foolproof and incapable of error.

And perhaps HAL was programmed to *want* to be in his new life aboard the *Discovery*. When he is interviewed by the BBC along with Bowman and Poole, HAL shows that he has a pleasurable rapport with his human counterparts:

> I enjoy working with people. I have a stimulating relationship with Dr. Poole and Dr. Bowman.

However, putting a 9000 computer in charge of a long-duration mission that the *Discovery* was on, was novel, since this was the first manned mission to the outer planets. So given this unusual situation, and despite his dialogue in the film, I cannot help imagining that HAL's transition went horribly wrong, and it drove him mad.

Notes

1. Similarly, do a search on YouTube and you will find countless videos of cats with amusing-enough behavior to be posted by their owners

2. Although the techniques for neural networks were identified nearly half a century ago, they only became common in recent years because computer processors became faster, at the same time that giant online databases of images became available to be used as training data.

Chapter 4: *Children of Men*

Clive Owen as Theo Faron walks by caged immigrants in *Children of Men* (Universal Pictures, 2006).

Alfonso Cuarón is the one of the most celebrated directors of our time. His breakthrough film, *Y tu mama Tambien* (2001), came as part of a Mexican Renaissance in filmmaking; Cuarón and directors Alejandro Iñárritu (*Birdman*, *The Revenant*) and Guillermo del Toro (*Pan's Labyrinth*, *The Shape of Water*) were dubbed the "Three Amigos" due to the fact they came out at roughly the same time. Cuarón helmed what many see as the best of the Harry Potter films, *The Prisoner of Azkaban* (2004), and then moved on to *Children of Men*. He would win the Academy Award for Best Director with his next two films: *Gravity* (2013) and the spectacular *Roma* (2018). He has also written, shot, and edited many of his own films. On *Children of Men*, he served as writer, director, and editor. Suffice to say he is a supreme talent and meticulous filmmaker in all aspects. As we have seen with Kubrick, those qualities translate well to sci-fi.

The film is based on the 1992 sci-fi novel of the same name by P.D. James. The title references a passage from the Bible: "Thou turnest man to destruction; and sayest, Return, ye children of men" (Psalm 90:3). We can take this to signify a great fall of men, very much in the biblical sense. That fall, and how far we've come in contemporary society, is exactly what is laid bare in the film. While the events are different from the book, and much of the narrative changed, the basic premise is the same for both. The film is set in 2027 England, where environmental disasters, war, and terrorism have destroyed much of the world. England is governed under a police—or Fascist—state with a strict immigration policy and tight security. Women have lost the ability to become pregnant (as my colleague Dr. Nicole Garneau notes in the second part of this chapter, this fact is changed from the book), and no new babies have been born in over 18 years. Faced with the end of civilization, the government offers a pill called "Quietus," which is essentially a suicide pill they distribute quit freely. In this mix, we find Theo (Clive Owen) who is thrust into the role of unwilling hero as he is tasked with delivering a (miraculously) pregnant woman named Kee (played by a wonderful Clare-Hope Ashitey) to a mythical organization who may be trying to save the society (although their existence is suspect and contentious). Theo's ex-wife Julian (Julianne Moore) tasks him with the job, and he is aided by old-friend and recluse Jasper (Michael Caine), an aging hippie who lives outside London.

The film was not a big box-office success, but it was critically acclaimed upon its release, mostly for its realist cinematography, its documentary-like feel, and its long takes that go on for sometimes uncomfortably long sequences. The film also skirts the line of sci-fi but ultimately includes many elements of sci-fi and certainly asks many of the central questions of sci-fi. It is also very much of the Earth. Since its release it has grown in importance as we now understand how so far ahead of its time the film was in the moment of its release. Screening the film in 2020, some 14 years later, and seeing the images of immigrants in cages and camps the size of small cities, was quite uncanny. Not to mention disturbing.

There are so many different contexts in which to read this film, and the critical analysis of the film since its debut has been rich and wide. Some of the commentary surrounds the innovative and realistic cinematography. Some has discussed the philosophical grounding of the film that is critical of the capitalist society. Some has discussed the dystopic not-so-distant future with its relevance to current-day politics. Still other commentary has placed the film into the mix of postmodern science fiction cinema. All of this commentary has credence and it highlights the complex and dense nature of the film, a film that certainly deserves all of the acclaim. The *mise-en-scène*, cinematography, editing, and sound all wonderfully point in the same direction, and the film has a consistency of purpose and focus that would make it one of the more influential sci-fi films of recent memory. It set the stage for the post 9-11 world of science fiction films that would last for over a decade, and the worldview therein would become a staple of the genre. A new staple.

One particular aspect of the film—namely, the vision of the future cities portrayed in the film—frames that worldview and allows us to view the film in terms of the post 9-11 world. One of those cities is London: a cross between the London of today, the New York of yesterday, and the Los Angeles of *Blade Runner*. The other city portrayed in the film, however, is the immigration camp, and while there is a rich reservoir of analysis in terms of the cinematography and the war-like conditions therein, I contend that it is here that the hopeful and humanistic messages of the film take root. It is here where the realism of the film—all of the cinematic aspects of the film really—meet up with its thematic project. The form and content not only set the stage for the larger trend in recent science fiction cinema of this period, but they also display how those cinematic aspects work together to situate the prominent tropes of sci-fi into a contemporary lens. That central question of sci-fi, "What does it mean to be human?," takes on a sociological lens that makes it not only real and timely, but most importantly, *urgent*.

That lens is important to consider from the fade-in and to continue pondering as the film moves along. Once again, the film tells the story of a future without children, where women have become unable to conceive and fertility is non-existent. The world is a mess: countries have been seemingly devastated by war, crime, and mass immigration; England has come under authoritarian rule; and the rest of the world may or may not have fallen into anarchy (we are never shown/told). Either way, there is a certain manic malaise portrayed in the film where the people of a dying, decaying society struggle to carry on. Theo starts out as a defeated figure in the middle of this *milieu*, content on drinking and gambling his way to avoidance. But the journey turns that hopelessness into a hopeful crusade and is a seemingly oppositional structure that plays out in interesting ways throughout the course of the film.

The cinematic aspects reflect this dichotomy. Certain elements of the *mise-en-scène* and cinematography display a realism akin to the best realist films in the history of cinema (Jean Renoir, Gillo Pontecorvo, or Vittorio De Sica, for example), with intricate camera, elaborate set-ups, and extreme long takes. But there are other sequences in the film where these elements become subsumed by the thematic project and a more reflexive and subjective atmosphere prevails. These rhetorical stylistics play out in the separate locations of London and the immigration city/camp. Both locations portrayed in the film become a literal and figurative battleground—with the ultimate prize being hope and future of humankind.

The portrayal of these cities is a study in opposites: the "civilized" city of London is one of strife, bombings, persecution, and alienation; the immigration city, however, while also torn apart by protest, fighting, and cruelty, at least offers a positive sign for the future—in the humanity and family we see inside the walls. It is just off the coast of this seaside city where that hope is realized, and a chance for re-generation and renewal is given breath. In the process, we are taken on that journey mostly through the character of Theo, who moves from that hopelessness (in London) to a more active agent for the future (in the immigration city).

The London of the film is a bifurcated city: one section where the majority live in disarray, confusion, apathy, and amid terrorism, and the other gated section where the upper-class lives in a mock Victorian-era time warp. The Victorian era is an apt symbol for this gated, apathetic society: the upper-class muddles on while society crashes around them. Emptiness and dis-connection are the only family interactions we see inside their homes. For example, when Theo journeys to see his cousin Nigel (Danny Huston), the Minister of Art, the family interaction consists of Nigel yelling at his son to take his pills. (His wife is not even there, but she sends her love, Nigel tells Theo.) The son, in a semi-catatonic state as he fondles a futuristic electronic device, follows orders without a word. At the same time, Nigel is entrusted with saving the most important art objects in the world (including Pink Floyd's floating pig, Michelangelo's "David" and a Picasso, among other things). The juxtaposition in this sequence is critical: the last remaining people on Earth have more interest in possessions than in interaction. There is no sense of personal or familial history.

Notice the cinematic aspects of the film here: The entire sequence is wallpapered with the British progressive rock song "In the Court of the Crimson King"; the cinematography highlights the colors of each segregated area—blues and grays for the masses and saturated colors for the gated community; the editing is of the continuous-classical Hollywood cinema variety; and the *mise-en-scène* highlights the paranoid, authoritarian, and apathetic nature of the society. Overall, it is reflexive and subjective. We aren't allowed many close-ups, and instead we see a fair amount of medium and long shots that place the characters in their environment, either blending in or in conflict with that environment. All of those shots add up to a sociological statement (rather than psychological/personal). It is society that is diseased.

By contrast, the immigration city we see later in the film displays a great deal of humanity and community in the midst of the absolute chaos on their streets. One particular sequence in the immigration city begins much like the visit to Nigel's house, where Theo moves from the crowds of London to his cousin's palace. In both sequences, we move from exterior to interior. This sequence in the camp begins as Theo and Kee—with new baby in tow after a profoundly unceremonious birth—are led through the teeming streets into an apartment. The journey is very similar to Theo's earlier journey, except in this instance, the camera lingers on the family pictures in the small apartment as the older women play with Kee and the baby. The contrast to the art of the Minister's house is striking—here, the pictures render a feeling of warmth and togetherness. Filmed in close-up, and in contrast to the earlier sequence, the cinematography affects this closeness and warmth. Coupled with the women playing with the baby as the men help Theo, the atmosphere is the antithesis of the earlier sequence at Nigel's. The move from the outer scenes of chaos in the street to the warmth of the apartment is Dante in reverse, whereas the earlier sequence in London is the opposite—the further you go inside, the more impersonal and distant it becomes; our descent into hell is wallpapered with impersonal "things" and possessions, the film tells us. The sequence in the camp warrants closer scrutiny, however, as the cinematic aspects co-mingle with theme quite nicely and work in stark contrast to the Nigel sequence.

Here, in the immigration camp, the specifics of the cinematic aspects are very different: the only sound is diegetic; the cinematography is predominantly hand-held, documentary-style camera; the long take is used in contrast to A-B-A-B style of continuity editing where A is the shot and B is the reverse-shot; and the *mise-en-scène* is a reflection of the real people with a real history—pictures, birthday candles, birds as pets. In a sense, these are two different films, and the stylistics define the rhetoric and drive home the themes of the film. In spite of the war and chaos in the camp, families live with pictures of their ancestors; ethnic groups form tight bonds; and acts of grace occur in the middle of the miscommunication, mayhem, and misery. It is hope that emerges from this milieu—hope for the future of men-kind. There are many different people who help Theo and Kee once they are inside the camp and after they become aware of Kee's pregnancy. The Romanian family even put their lives on the line to help Theo and Kee escape "The Fishes" (the radical group chasing Kee and wishing to use her as a political pawn). Some die in the process, but ultimately, they are able to help Theo and Kee get into the boat and make it out to sea. A specific example of the mad *milieu* from this sequence is when Theo and Kee take the baby out of a crumbling building in the middle of the long and intense firefight between The Fishes and the government. In a striking scene, everyone stops shooting for a minute, but once the baby leaves, they quickly resume fighting. The point: they are more interested in fighting than they are in rebuilding humanity; they have forgotten what they are fighting for.

The people helping Kee and Theo see through this oversimplified opposition and come down on the side of humanity. In the process, the immigration city becomes the only location in the film with any sense of hope, family, and history. And it is in that city that life is reborn—literally and metaphorically.

The much-discussed realism of the film is the aesthetic counterpart to this theme of rejuvenation. The natural lighting and hand-held camera of the immigration city stand in stark contrast to the reflexive, stylized aesthetic of London and especially of the Nigel sequence. The realism inside the camp, in other words, lends an authenticity to the immigration city that speaks to the film's themes of hopefulness and humanism.

Aside from these thematic dimensions, however, the aesthetic aspects of the film speak to the larger science fiction tropes of the past fifty years—a lineage of which *Children of Men* is part of a particularly promising progeny. I have always seen *The Matrix* as one of the key films in the genre and it marks a turning point in contemporary science fiction cinema. When Director Darren Aronofsky (*Requiem for a Dream, Pi, Black Swan, The Fountain, Mother!*) went to see *The Matrix* for the first time, he commented that it was the death of science fiction cinema as we knew it. He was referring to how *The Matrix* "took all the great sci-fi ideas of the 20th century and rolled them into a delicious pop culture sandwich that everyone on the planet devoured. Suddenly Philip K. Dick's ideas no longer seemed that fresh. Cyberpunk? Done." Despite the obvious hyperbole here, Aronofsky has a point. To extrapolate, I see science fiction moving toward a new, humanistic focus with less of the fetishistic technophilia (the fetishizing of technology) so prevalent in films leading up to and including *The Matrix*.

Four examples from the post–*Matrix* era help define this new thread, including Steven Soderbergh's *Solaris* (2002), *Children of Men* (2006), *Moon* (Duncan Jones, 2009) and Aronofsky's own *The Fountain* (2006). The four films are aesthetically different yet thematically similar: there is a decided lack of computer-generated-imagery (CGI) in favor of realism, metaphysical themes, absence of technophilia, and most significant of all, a return to the humanism of the early days of science fiction cinema and a return to the essential questions of science fiction. This new strain of humanism speaks to the changing nature of science fiction cinema as it simultaneously reflects and comments upon the contemporaneous societies and, even, our current era.

The brutal realism of the not-so-distant-future in *Children of Men* relates directly to the brutal realism of our filmic past and present—the immigration city sequence is a direct descendent of landmark films such as Pontecorvo's *The Battle of Algiers* (1966), or even Steven Spielberg's *Saving Private Ryan* (1998). This is now a tangible future that is felt bodily and immediately, bridging the distance between a far-away future that is safely out of reach for us, and our own present-day reality. *Children of Men* foregrounds such concerns through its authenticity. The end result of this realism is the questioning of humanity's actions in the present: How do we remain human when faced with the end of humanity? How do find grace in a world seemingly void of any spiritualism? How do we help one another when we can't help ourselves?

These questions move science fiction away from that safe, far-away future of the fantastic disasters of '40s/'50s/'60s sci-fi and place it firmly in our grasp—a placement that forces us to think not about the relationship we have to technology in a distant future but the relationships we have with each other, in our world, today. Science fiction has always been good about making social comments about our current society but placed in a distant future, rendering it all easier for us to digest. This film changes that dynamic—it places societal ills into a near future, and the result is a harsher warning. What are we warned about? Immigration, and our attitudes toward it. The environment, and our neglect (and outright plunder) of the natural world. Apathy, and our current attitude toward each other and toward our environment. Again, considering the world of 2006 and the world of 2021, it almost seems as if the film was produced today and sent back in a time-machine to 2006.

The subject matter is now immediate and personal rather than distant and impersonal. In *Children of Men*, characters struggle with how to salvage their remaining humanity or if they should salvage it at all. Throughout the film, Theo remains indifferent to humanity's plight and seems content to drink the rest of his life away. His wife Julian, however, still fights an underground battle against the government and helps Kee find safe passage and safe harbor—with Theo. Many people assist Kee and Theo along the way: they receive help getting into the immigration camp and once they arrive there as well. This grace seems to change Theo on the journey; in the end he winds up putting his life on the line for the future of mankind—helping Kee through the insanity of the immigration camp to reach her destination. Theo finally chooses grace and life over the selfless concerns shown by many others in the film. Yet the sheer amount of people in the film who do put their lives in jeopardy for the benefit of mankind speaks to the film's humanism. In a world where prejudice and selflessness seem to rule—where the government can preserve artifacts while persecuting immigrants and forcing people to commit quiet suicide—the characters choose humanity. That central question of science fiction, "What does it mean to be human?" is examined on both a personal level and on the societal level. The aesthetic of the film as well of themes of the film thus point to a new direction in science-fiction film, one that would play out in interesting, wonderful, and stylistically varied ways over the next decade. *Children of Men* starts this trend.

—Vincent Piturro

Infertility and the Near-Future

Nicole L. Garneau

The disdain on Theo's face is palpable as he wades through the crowds at the café. Who are these people, mourning the death of a young man they didn't know, as if the child was in fact their own? Baby Diego was the youngest human, the last to be born before the epidemic of infertility effectively stole children's laughter from the planet. And Theo, played by Clive Owen, is still mourning the death of his own son, 20 years later, as he watches this egregious show of international mourning with disgust. Moments after leaving the café, we see a couple embracing in the background behind Theo, the briefest moment of humanity. Then bang! The café explodes into fire, and Theo's near-death experience leaves him with a ringing in his ear and further repulsion for what the world has become. "Woke up, felt like shit. Went to work, felt like shit." To which his good friend, Jasper, played by Michael Caine, replies, "That's called a hangover." In retort, Theo expounds, "No, at least with a hangover I feel something."

Infertility thus sets the scientific backdrop for some of major subjects and themes the film approaches: destruction of the natural world, xenophobia, and complacency in the face of moral crisis. Unlike many science fiction films that take on the idea of the human health condition through the lens of a wholly fictional epidemic, *Children of Men* is unique in that the health condition, like the setting itself, is near sci-fi. In turn, there are precious few conjectures to make in order to understand the science, because we are living it now. For this reason, and the sheer terrifying nature of its truth, this film possibly more relevant now than it was in 2006 when it was released.

In order to understand infertility, we need to head back to our middle school health ed class to review the process of conception. To paraphrase the Center for Disease Control and Prevention (CDC), to get pregnant, a woman's body must release an egg from one of her ovaries (ovulation), and a man's sperm must join with the egg along the way (fertilization). The fertilized egg must travel through the fallopian tube toward the uterus (embryo transportation), where it must attach to the inside of the womb (implantation).

Infertility may result from a problem with any or several of these steps. Below, each step is described, alongside possible contributors to infertility.

Ovulation, in an otherwise healthy woman of child-bearing age, can be impaired by biological factors (increased testosterone, diminishing egg reserves, changes in brain hormones), lifestyle (obesity, excessive exercise, stress, or low body weight), and premature menopause due to exposure to medical treatments such as chemotherapy and radiation. Problems with fertilization are primarily due to abnormalities of the sperm, specifically the number of sperm (concentration), the shape of the sperm (morphology), and how well they swim (motility). Like in egg release, sperm production and release can be affected by biological factors (genetics, large testicular veins/overheated testicles, changes in brain hormones), diseases (diabetes, cystic fibrosis, celiac disease, certain types of autoimmune disorders), lifestyle (nutrition, heavy alcohol use, obesity, smoking, anabolic steroid use, illicit drug use and certain vitamins and supplements), cancer treatments, and possibly heat exposure to scrotum. Embryo transportation can be derailed in women due to many diseases, from appendicitis to STDs to endometriosis. And finally, successful implantation depends on the environment of the womb. An infertility issue more common among overweight individuals and women of color occurs when the embryo cannot implant due to the presence of benign tumors in the uterus called fibroids.

This complexity, from the seemingly most basic of human functions of conception, makes understanding, diagnosing, and tracking infertility one of the more difficult of all human health concerns, and therefore, one of the more devastating of possible epidemics. The problem is escalated when we consider that there is mounting evidence that a possible underlying cause of both male and female infertility, although the mechanism is yet undiscovered, may be environmental contaminants.

It is not explicit in the film which came first, infertility or industrialization, that led to a world-wide climate crisis. However, according to Theo, "Even if they discovered the cure for infertility, it doesn't matter. Too late. The world went to shit. You know what? It was too late before the infertility thing happened." Theo's take may be nihilistic but also accurate. Industrialization in this film is linked tightly with the apocalyptic images of the cannibalism of the natural world. The consequences of this man-made destruction is destabilizing to macro climates, thus placing an unfair burden on populations in the more vulnerable locations where environmental changes have the most devastating effects. Then, the lack of resources, opportunity, and ability to care for one's loved ones in such places leads to, as it always has, mass migration and the consequences that follow such migrants then stem from xenophobia in their new homes.

According to the World Health Organization, it is difficult to expressly pinpoint infertility prevalence rates. It is much easier to track fertility rates (the average number of children a woman gives birth to in a lifetime). However, fertility rates are masked by many other factors, including the following, according to the 2017 global health metrics published by *The Lancet* in November of that year: fewer deaths in childhood, contraception, an increase in the education of women, and more women working outside the home. Although fertility rate is not a perfect proxy statistic in relation to actual infertility (defined by the Centers of Disease Control and Prevention as not being able to get pregnant [conceive] after one year [or longer] of unprotected sex), what it does show is telling. We see that in West, East, and Central Africa there remain some of the highest fertility rates in the world. In the film, this is aptly represented by Kee; she is an 8-month pregnant "fugee" in the developed country of England, and she is of African lineage. Going back to the attempt at determining infertility rates, there is an interesting correlation between infertility in less industrialized nations; however the burden in this case is due to neither male nor female factors, and in reality, is due to infectious diseases. All told, the data leads one to conclude, as it relates to this film and our possible future, that people in less developed countries are more suited to withstand a possible epidemic of infertility.

Diving further into the science of infertility, it is of note that Cuarón elected to change a key infertility element when he adapted the book into film. The director places the burden of infertility solely on women as Jasper calls it, during the telling of a joke, "the ultimate mystery, why are women infertile?" In contrast, the CDC states that, "Overall, one-third of infertility cases are caused by male reproductive issues, one-third by female reproductive issues, and one-third by both male and female reproductive issues or by unknown factors." Although more scientific data estimates that infertility is actually more like 40–50 percent due to male factors. Moreover, meta-analysis studies have put forth evidence to confirm that a global decrease in sperm concentration (50 percent over the past 5 decades), aligns with both the decrease in fertility rates and the estimated increase in infertility rates. Finally, there is recent data to suggest that the environmental conditions/contaminants and nutritional access/choices of our parents and grandparents can be passed down through epigenetics, meaning that through no fault of our own choices, there are markers on our very DNA that survive generation after generation, that cause infertility.

The impact of infertility, as shown both in the film and by sounding alarms of many scientists across the world, is inconceivable (pun not intended!). A review of infertility trends published in 2015 beautifully relates the concern we should have: "Infertility is a condition with psychological, economic, medical implications resulting in trauma, stress, particularly in a social set-up like ours, with a strong emphasis on child-bearing."

Kee's midwife, Mariam, played by actor Pam Ferris, explains it this way, "As the sounds of playgrounds faded, the despair set in. It's very odd what happens in a world without children's voices." Metadata analysis of fertile and infertile couples show an increase in depression in infertile couples and an increase in severe depression in infertile women. When we align this idea with the harsh reality of changes in the micro-climate of industrialized cities, as shown by the pervasive grey tones utilized in the film's city-based scenes, we understand the literal toxicity of things like air quality, but also figurative toxicity in how it affects our mental health. In this way, Cuarón was not far off to have his fictional English government supply the masses with antidepressants and assisted suicide kits in their rations. The depressed psychological state of the populace in this film is compounded with the environmental stress on the planet (as a cause or correlation), further dividing the have and have nots, it sets the stage for mankind resorting to untold immoral acts on the way to the extinction of the species. This is represented starkly in the film, but also in other recent dystopian works of fiction like *The Handmaid's Tale*, by Margaret Atwood (and its wonderful TV adaptation). In an argument on how to best move forward to save Kee and her baby, Theo's suggestion to make it public is denounced: "We all know the government would never acknowledge the first human to be born in 18 years from a fugee."

With the depth of the science and social commentary at the forefront of our minds, let's close with the interaction between Theo and his elitist and privileged cousin, Nigel, played by actor Danny Huston. As they stand among the greatest works of art, salvaged from the world's disintegrating countries, including the recently salvaged statue of Michelangelo's "David," Theo just smirks. "What?" asks his cousin. "You kill me." Theo replies, shaking his head. "A hundred years from now there won't be one sad fuck to look at all of this. What keeps you going?"

"You know that it is, Theo? I just don't think about it."

It seems Nigel was exactly the type of person Dante had in mind when he said, "The darkest places in hell are reserved for those who maintain their neutrality in times of moral crisis." And while this near sci-fi look at the human condition is in many ways a warning, it also leaves us with hope for the planet. It is unclear if Kee and her baby will survive, and if that survival will be the miracle in which the players in this film place their hope. What is clear is that nature is reclaiming the planet. The sadly abandoned schoolhouse where Kee, Mariam and Theo await their escort to break into the refugee camp is being taken back by the Earth by plants growing, water dripping, and a deer passing through its halls. It is clear the human species as we know it may not survive, but life itself will rebound, as it always has and always will.

Chapter 5: *Perfect Sense*

Eva Green as Susan and Ewan McGregor as Michael in an eerily prescient scene from *Perfect Sense* (BBC Films, 2016).

David MacKenzie has directed 10 feature films since 2002. His first film of note was *Young Adam* (2003), starring Ewan McGregor, and he followed that up with *Asylum* (2005) and *Hallam Foe* (2007). His biggest-budget film and most noteworthy was certainly *Hell or High Water* (2016) with Chris Pine, Ben Foster, and Jeff Bridges. That film would gain attention in the awards circuit worldwide, including four Academy Award nominations: Best Picture, Original Screenplay, Best Film Editing, and Best Actor in a Supporting Role. While it did not win anything at the Oscars, there was some talk about the film being a dark horse that year. *Moonlight* (2016) eventually won the award in the now infamous ending to the ceremony where *La La Land* (2016) was initially announced as the winner. *Arrival* was also nominated that year, in somewhat of a coup for science fiction, and with *Hell or High Water* also nominated, classical Hollywood cinema genres (science fiction and the Western) made a comeback. But before he made his foray into the grandest of all the old genres, MacKenzie made *Perfect Sense* in 2011, an atmospheric and realist gem of a science fiction film that is more independent film than big-budget Hollywood, albeit with two A-list actors. The story concerns two self-admitted "assholes" who seem completely self-centered and self-important at the start of the film, only to find each other amidst a world that is fracturing by the day. That fracturing is the film's central conceit—the slow removal of our senses on a mass scale that leaves everyone without smell, then taste, and so on until all of our sense are gone. The film doesn't deign to tell us why this happens—although we receive generalized hints—but rather the focus remains on the effects of the phenomenon, how the world copes on a large scale, and more importantly and succinctly, how we cope as individuals. In a world without senses.

The film fits into a larger trend that I have referred to as near sci-fi, which became more prevalent and increasingly dense in the post–9/11 era. As previously discussed, the pre–9/11 era saw a trend toward highly formalist, CGI-infused master narratives about AI, virtual reality, and the idea of a world imposed upon us. The trend fits into the larger scheme of sci-fi in the context of time. If we look at the history of science fiction, we can see consistency in the overarching subjects and themes: the '20s/'30s dealt with Utopias and Dystopias (machine age, industrialization); the '40s/'50s were very much a function of the post–World War II zeitgeist as well as the various sightings reported around the world, reflecting a growing interest in UFOs, aliens, monsters but also coinciding with the increasing technical proficiency of film; the '60s looked at space exploration and extrapolated from the contemporaneous space race; the '70s branched out into sub-genres and expanded the breadth of sci-fi as it delved into the relevant topics of the period and thereby reflected the changing world, such as environmental concerns, changing ideologies (race relations, feminism), and the burgeoning chokehold of corporatization; the '80s and '90s saw a turn toward developing AI, the fast-moving takeover of technology, and the idea that our world may be no more than a computer program in which we are all just binary code run by machines. The films of those eras not only infused these subjects into their narratives, but the attached themes also amplified the cultural *zeitgeist* in terms of the desires, obsessions, and fears of the various eras. If we take a broad stroke look at the history of sci-fi from the '20s to the turn of the century, we see a clear pattern: the initial fears of industrialization turned into a fear of the Other which then turned into a fear of machines. Drilling down into more recent sci-fi, the human vs. machine theme that is laid bare in *2001: A Space Odyssey* continues on through *The*

Matrix. We have been on this trajectory since Kubrick's masterpiece, where we saw the first fight between humans and machines, with the humans eventually winning there. After *The Matrix*, where the power dynamic is reversed and the machines prevail, there is no longer a subject and object—man and machine are now both subject *and* object. The primary oppositional sci-fi structure of the period was shattered. Then, 9/11 changed everything.

The event of 9/11 provided the seminal grounding and triggered the new trend toward what I call "Near-Sci-Fi." The new, primary oppositional structure would be humans vs. humans, and the themes would wring out the various ways in which we don't need an "Other," alien or otherwise, to lead us to imminent doom. The only thing we need is *us*. The films thus reflect this inward turn: *Children of Men*, as discussed in the previous chapter, may be the most important and the seminal film of the movement, crashing into subjects such as immigration, overpopulation, environmental disaster, and widescale genetic disorder while giving us a realist aesthetic that shouts "urgency of now." In other words, *Children of Men* places a giant mirror onto the world of the period, and then other films would follow along. *Wall-E* (2008), *Moon* (2009), *The Road* (2009), *Monsters* (2010), *The Book of Eli* (2010), and *Snowpiercer* (2011), among others, all follow this path. *Perfect Sense* thus fits right into this larger trend, with its laser focus and its realist aesthetic that makes us feel as if all of this could happen right now.

Director David MacKenzie addressed the sci-fi aspects of the film in an interview at Sundance:

> Well, there's only a small … the sci-fi element I guess is tied to the present. Do you know what I mean? There's a sense that it could be a sort of plausible near future as opposed to something further away. I didn't—actually I'm about to start writing another sci-fi script so I'm probably, by myself, drawn to that. But I think that there's something about the way that our time is right now where we are aware of the finite resources of the Earth and the technology is getting faster and faster. The near future and this future seem to be kind of merging with general narratives. And I think there's something attractive about that at this point.

MacKenzie's vision is frightening, sobering, and strangely, in the end, uplifting, giving us yet another crack at the central question of science fiction: "What does it mean to be human?"

We meet the two principals as the film opens, doing what they do and showing who/what they are: Susan (Eva Green) is first seen in long shot, walking with her sister along a barren seaside landscape of sticks and seagulls, plodding along in the mud. The long shot and long take are the essential elements of film realism and thus the film announces itself as thus. We learn little of Susan from the dialogue, other than she has issues with men, or rather, keeping one around for long. The other side of the same coin is Michael (Ewan McGregor) who we meet shooing away a woman from his bed because "he can't sleep while she is there." We soon find out that each has a problem with human connection. Michael had a sick girlfriend whom he left when she got sick; he says he used to visit her grave to make himself feel guilty, but now he doesn't visit much anymore and feels less and less guilty. He fears he may have lost the ability to feel—a prophetic sentiment. Susan relates that she can't have children and she and her former *fiancé* split up not very long ago. He recently married, and his new wife is pregnant. Right from the start, both of our main characters are broken figures in a seemingly broken world. That world is one planted firmly in the here and now.

Michael heads off to work on his bicycle, and we get a shot of him that Darren Aronofsky called a "slurry cam" when he used it in his own film, *Requiem for a Dream* (and first used by Martin Scorsese when he attached it to a drunk Harvey Keitel in *Mean Streets* [1972]). The camera is attached to his bike, looking up (closely) at Michael and thus following along as if we, the viewer, were the handlebars. The jittery, disorienting, documentary-style nature of the shot places us into the action, making us part of it and immediately engaging our sense of movement/motion. Both of these opening sequences where we meet our protagonists serve to immerse us into their reality, at once giving us a sense of being there as well as allowing us to feel what they feel and sit with them as they live. These sequences also reveal the spectrum of realism in film: one is very moderated—the long take and the long shot—and allows us to make our own decisions about where to turn our attention and our feelings. The other—the documentary-style realism of the slurry cam and hand-held camera in this case—makes us feel as if we are moving along with the subject and along for the ride, as though it were a documentary film and the action is real. Both achieve a similar effect of placing us closer to the action and allowing us to enter the world as we know it and feel it.

The *mise-en-scène* continues its argument that what we are seeing is of the here and now, and this is not a faraway future. Very much like many of the other chapters in this book—*Arrival, Interstellar, Children of Men*, and *Upstream Color*—we are not assaulted with a future world of fetishistic technophilia or a plethora of possibilities. We are assaulted with an abundance of actuality; a world this is very much ours, yet one that is fraying, diseased, and/or decaying. The cityscape of *Perfect Sense* also adds to the urgency of the narrative: it is not futuristic or noteworthy of anything contemporary; in fact, it is an older city that has a lived-in/used feel to it. And while it is not named, it is pretty obviously Glasgow, Scotland—hardly a bastion of futurism (with all due respect to the wonderful city and its inhabitants). It is wonderfully *used*, and it is certainly of the Earth.

The editing in this comparative opening speaks to one of the main subjects of the film and of all science fiction: connection; the attached theme is very clearly the need, and even intense desire for, humans to connect with one another. While the dialogue and action speak to the inability of our main characters to connect in any meaningful way, the editing parallels the two characters and therefore compels them to do that very thing they are unable to do: connect. In other words, the filmic apparatus—the editing— is our (the viewer's) stand-in and does the work of the connecting for us. We see that these seemingly disparate characters are unable to connect and thus, we see their similarities and their supposed shortcomings. Aside from being wonderfully human and terribly common, this inability to connect, we also see that such a shortcoming is what will bring them together and help them to find each other. The editing therefore serves to connect the two characters and allow them their own special brand of humanity that most of us understand.

Soon after, they meet. Their meeting coincides with the first mass loss of sense—smell. As an epidemiologist, Susan is made aware of many such cases around Europe and soon after, the rest of the world. Each loss of sense is always preceded by a spontaneous emotional outburst, as if the afflicted were experiencing a loss and then grieving in the same moment. Susan experiences this outburst soon after they meet, as Michael is cooking for her, and he consoles her. They retire to her apartment and we cut to Michael waking up next to her. He experiences his own outburst and she, in turn, consoles him. The next morning, they have both lost their sense of smell. Their meeting thus coincides with a loss, giving them a shared experience and then an opportunity to make new ones—a connection they have both been failing to make.

We also get our first montage interlude of the film at this point: each time the world loses a sense, the film injects a montage of images coupled with an omniscient voiceover describing what is happening/what is lost/ how the world deals with/why it is happening. These montage sequences give us clues as to the genesis of the events, but, and a warning here to those among us who "have to know WHY it's going on," the film never tells us exactly why any of this is happening nor does it seem to care. These interludes are emotional, interstitial interludes that are highly formalist (taking us out of the realism of the film and imposing the film on us as a work of art, rather than an event happening in the here and now). They are the filmic equivalent of the emotional outbursts of the characters. The montages serve to highlight the shared loss and the shared grieving, yet at the same time, the shared perseverance and shared fortitude of the world as everyone moves forward. For example, the first montage after smell disappears begins with the words (in voiceover): "Life goes on."

As the montage makes clear, life going on is nothing short of a miracle. Michael is a chef, and his restaurant adapts and keeps going; as the montage begins, they are not sure they can stay open and stay in business, but they give it a try. They use more spice, more salt, more of everything; they adjust to the senses we still have and cater to them. People adjust as well, with the hardest part being the loss of memories and experiences triggered by the smell: "Cinnamon might have reminded you of your grandmother's apron.... Diesel oil might bring back memories of your first ferry crossing...." The dialogue ends but the visuals go on, driving home the point of lost memories. The quick cuts of these visuals also remind that editing, in film, can mirror the memory process—the quick cuts and short bursts mimic our process of remembering; it is a beautiful mix of theme and content. The montage ends on a positive, high note as Michael and the restaurant owner smile at their newfound hope. Life goes on.

Following the montage, Susan and Michael meet again. They take a walk. They happen across a street performer in the middle of an act. Her act is meant to help people remember how to smell—the artist tells the story of a walk in the woods and uses the other senses to conjure up smell. She touches a leaf to Susan's face, and she tells of the sounds of rain and the feel of moss under your feet. "Enjoy the air. Enjoy the moment," says the artist as she begins to play the violin and adds a "soundtrack" that helps everyone make new memories. The message here is wonderful: art helps us to feel, helps us to remember, helps us to connect. Movies, for example, can stand in for memories. Music can stand in for a formative experience. A painting can stand in for an awakening. A poem for an emotion. It is at once synecdoche and metaphor. This wonderful moment is made stronger by its connection to the theme of the human drive for connection—Michael and Susan experience this together, and the artist holds court with a group who all experience it together as well. For Michael and Susan, it not only serves to help them resuscitate their sense of smell, but they are creating new memories, together, that become formative experiences for them—connected to emotions. The cycle of connectivity is made apparent in this one performance as life goes on, hopefully and even positively.

In an interview at Sundance, reviewer Dan Mecca posited that what makes the film "risky in its own way is how positive it is." In response, MacKenzie stated "And I don't know how much of that is true, but people are constantly confronted with adaptations and possibilities and deal with it without even thinking about it. The idea of taking that as a sort of an optimistic, human positive thing seemed to be a really beautiful way of expressing the humanity and the magic of humanity." That optimism about humanity—and what it means to be human—continues throughout the film as every sense gives way and people adapt. When everyone begins to lose their hearing, for example, sign language becomes prominent and Susan's lab even posts boards with sign language instructions. Customers eat at Michael's restaurant and communicate without missing a beat, enjoying meals and each other's company. Life goes on.

The film also allows us, the viewer, to feel what the characters feel at this point. As the characters lose their hearing, WE lose our hearing. The diegetic sound is lost (that sound emanating from the world of the film), and their silence is our silence. Our loss is their loss. Once again, the realism of the film drags us in, and it not only allows us to sympathize with the characters, but it implicates us. What will we do with this knowledge?

We also get another montage that precedes this loss, with possible reasons for the entire malady. But the reasons point to almost anything: politics, God, a polluted environment, nefarious and/or rogue state actors, genetically modified plants, etc. The sum total adds up to the fact that all of these actions are made by humans and harm humans in some ways. The only remedy, it seems, is something we are told later when everyone is about to lose their sight. As the montage states: "People prepare for the worst but hope for the best. They concentrate on what's important." The characters go outside, look at the sun, they cherish pictures of loved ones, or they simply enjoy each other. Again, even with the possible loss of sight looming, the message is positive, and the characters find the good and the hopeful. We are further implicated—what would do then and why are we waiting? What it means to be human is quite evident throughout the film.

The final sequence of the film poetically sums this all up and finds Susan and Michael frantically trying to find each other as their loss of sight looms over them. They both go to each other's houses, missing each other as they do. Yet they still manage to come together and embrace. The film goes dark, and the voiceover tells us they still grope for each other in the dark and still feel each other. They can't hear, they can't see, they can't smell, they can't taste. But they still connect. And life goes on.

—Vincent Piturro

All They Need to Know: Connectedness

Nicole L. Garneau

Above all, *Perfect Sense* is a story of connectedness amidst an apocalyptic backdrop. The scientific subthemes of the film, including epidemiology, emotions, sensory perception, and adaptation, are the threads that weave together this unique take on the genre. In exploring the science behind the science fiction, we can take an introspective look at our own modern day survival, the connectedness we take for granted and dismiss for the sake of progress, and finally, Would we still be able to survive should we completely lose our ability to perceive and connect, and therefore control our world?

Connection, Contact, Pattern

Epidemiology is the study of a health concern from a very holistic perspective. Epidemiologists are like disease detectives. They try to understand not only what the disease is, but where the disease is, who has it, and how it spreads. The goal is to use this information to figure out how it all is connected in order to control it from spreading further. The female lead in *Perfect Sense*, Susan (played by Eva Green), is an epidemiologist that is specially called in when the first cases of sensory loss, specifically overwhelming grief followed by the loss of smell, are identified as a possible epidemic. When she asks how the victims are infected, her boss responds that there is "no connection, no contact and no pattern, so we won't panic." Reports of mass numbers of people afflicted globally pour in, it's now clear that they have a pandemic spread of disease and are still at a loss as to the cause. "It's fair to say that it's not infectious … it's fair to say it's spreading."

The First Symptoms: Emotion to Sensory Loss

Many people think of emotions and sensory perception as distinct and separate neurological processes, and in fact that's how most scientists study the two topics, which is to say entirely separately from one another. In *Perfect Sense*, we have the pairing of a distinct emotion with the loss of a distinct sense. It begins with what the film's scientists call "Severe Olfactory Syndrome" or SOS. Extreme grief is followed by a loss of the ability to smell. There is strong evidence for why the film's writer, Kim Fupz Aakeson, linked these two processes. According to an article published in the journal *Emotion*, "emotional and olfactory processing is frequently shown to be closely linked both anatomically and functionally." Which is to say that your olfactory bulbs, located almost right behind the inner corners of your eyebrows, not only are physically located right next to the part of your brain that helps you process emotion, feel and react (called the limbic system), but they are also the only area of the brain that sends one way signals to your amygdala. The amygdala is a tiny area of the brain with a hugely impactful role in our lives: it helps us coordinate responses to the things we see, feel, hear, taste, and touch in the environment, especially those that are emotionally charged. Many of us are familiar with the eerie and wonderful sensation of reliving a memory when you experience a familiar odor from your past. It is no coincidence that these aromas are so strongly linked with vivid memories. When we smell something tied to an emotional experience, the brain cells that hold that well-preserved memory are directly linked to the sensory neurons that hold the related smells that were there when that memory was formed. This linking leads to these neurons firing together when we smell that familiar aroma again, and the result is that we remember.

The only documented causes of sudden loss of smell and/or taste function is head trauma (less likely with taste loss), a cerebrovascular accident, acute upper respiratory infection, surgeries that could damage cranial nerves, and lastly, some psychiatric conditions. So while we don't have data to form a strong conclusion for why the characters experience a sudden and complete loss of olfaction after the emotional exhaustion of grief or the loss of taste following fear, there is strong evidence to link how each emotion is connected to each respective sense.

First let's look at olfaction and sadness. It is difficult to assess cause and effect, however there is much evidence to support a correlation between depression and feelings of sadness to a decrease or lack of an ability to smell. Second, we consider taste. Acute stress leads to increased heart rate and blood pressure and leads us to be able to taste things more strongly at that moment, and therefore, following stress, we return back to a baseline ability to taste. In the case of *Perfect Sense*, when the characters experience intense fear and terror, perhaps it shocks their system so much that when it ends, there is a complete loss of taste. A loss of taste when paired to a loss of smell doesn't explain the hedonistic hunger and the delusional devouring of edible and nonedible things alike, but it does make it impossible for victims to rely on signals from the nose and the mouth to tell the brain that they are consuming something potentially dangerous. It is only when their clarity returns that they can lean on mouthfeel cues, sight, sound and memory to shock them into realizing they have eaten things that could harm them, and therefore feel disgust.

The ability to smell is a powerful thing, and above all other senses, it can bring us back to a time long passed. Losing the ability, in addition to dampening our ability to recall certain memories in clarity, also decreases our ability to sense dangerous poisons and toxins, experience pleasure from eating and drinking, and inability to smell the scent of a loved one. These side effects are then compounded when one loses one's sense of taste, negatively impacting the quality of life.

The film then mentions a baby born with all her senses intact. Once more there is hope, "might be a chance for antibodies." There are two well-known cases and causes of natural immunity to infectious diseases. The first is smallpox. An English doctor by the name of Edward Jenner realized that milkmaids were naturally immune to the disease. The milkmaids had been exposed to a similar virus, cowpox, and had built up antibodies against the disease agent. Because cowpox and smallpox were so similar, the antibodies they developed against cowpox also protected them against smallpox. Jenner then successfully tested his theory by "vaccinating" people using the pus from cowpox sores on the milkmaids' hands. The second example is HIV/AIDS. There is a group of people that have been exposed to the virus repeatedly, but they have not become infected with HIV and have not developed AIDS. The reason is genetic. They were born with a DNA mutation for a specific cell receptor (like a lock). This mutation changed the shape of the receptor lock, and because of the change the virus can no longer attach and unlock the cell in order to infect it.

This history is likely what Susan and her colleague have in mind as they hope to determine if there are people, like this baby, that are naturally resistant to the sensory loss. If so, they can harness the antibodies to develop a vaccine or possibly even a cure. From this point we see the characters continue living. No one is quarantined, but no one is sure. Life returns to its normal cadence.

I Think It's OK to Panic Now

Rage, hatred, and unchecked anger and violence lead to "severe hearing loss syndrome." Negative emotion has been shown to make sounds appear louder. Again, as we saw with taste loss, perhaps the extreme expression of negative emotion led to such an intense perception of sound that when the emotion collapsed, so too did the sense of hearing.

According to the National Institute on Deafness and Other Communication Disorders, sudden deafness in one or both ears is considered a medical emergency. Patients often become dizzy and may experience a ringing in their ears. In real life 50 percent of people recover spontaneously and 85 percent who receive treatment from a doctor will recover some of their hearing. Some of the more common causes of sudden deafness are infectious diseases, head trauma, and blood circulation problems.

Like severe olfactory syndrome, the spread of severe hearing loss syndrome is distinct (unlike taste which appeared to affect everyone at once). For this reason, perhaps, we see people being quarantined under house arrest for the first time. People have time to contemplate the loss of sound, and we experience alongside the main characters, Susan and Michael (played by Ewan McGregor), the cacophony of sounds and the appreciation for bells and birds. We are also devastated by the sounds of screams and babies crying. Michael is the first to succumb, and Susan, in fear for her safety, flees. She succumbs next, throwing the phone to Michael, and says, "I love you" before kicking off her own tirade with the words, "What disease! Where are you hiding?" and then repeats the word "why, why, why, why" in her extreme frustration and anger.

People soon return again to a sense of normalcy. There is hope in communication, in the aesthetics of meals, if not the taste of sound. There is time to take in the beauty of things important to them, and they take advantage. Equally so, people prepare for vision loss.

Sudden vision loss can be due to a blockage of light entering or passing through the eye or damage to the optic nerve. The most common causes are due to a block in circulation to the eye structure, blood in the back of the eye, or eye injury. The loss of sight in the film is described by the narrator as analogous to the ice age. People view the ice age as gradual, but scientists have found mammoths with undigested grass in their guts. Perhaps the cold came quickly, taking them all off guard. These words segue to an avalanche and the narrator foretells, "That's how the darkness descends upon the world."

But first there is joy: the sense of joy and happiness, love and gratitude, of what it means to be alive. The connection between love and vision is the most uplifting of any of the pairings and is backed by research in cognitive science. Our emotions have the ability to change our perception of the world. Called the "affect-as-information" hypothesis, it looks at how mood and emotions change judgment, attention, and thought. As a flood of positive feelings wash over each character, we see them reach out for human connection and embrace and genuinely take in how interconnected we all are. Michael and Susan are no exception. They seek out one another, their emotions guiding them to see the value in love and setting aside their vulnerability to experience love. They embrace as the darkness washes over, leaving them only with the sense of touch to survive.

How They Adapt: Beyond Fat and Flour

At each stage of the film and with each loss of a sense, we see adaptation. It is this theme of adaptation in which Michael shines as a counterpoint to the seriousness of Susan. In one of their first scenes she self-describes her career as "death and misery" to which Michael replies, "Ok, I'll get something sweet. Sweet always makes you feel better." Michael is our conduit to consider the connection of perception and survival to pleasure.

Chapter 5: *Perfect Sense*

Our senses are our conduit to the world. According to Steven Frigs in the book *Sensory Perception: Mind & Matter*, "The sensory organs we have today are the result of millions of years of adaptation to the needs of animal in their struggle to survive ... all sensory systems fulfill a clear and vital purpose: the survival of an individual and the continuation of a species." We sense sights, sounds, tastes, smells, temperature, textures, and pain. These clues from the environment are sent from our sensory organs to our brains. That tiny part of our brain that serves as our emotional center, the amygdala, kicks an immediate physiological reaction into gear. The result is that our body does something immediately, unconsciously, to protect us in that moment. Then rational integration begins to occur. What we sense is analyzed with more thought, incorporating what we've learned previously and our memory of outcomes in similar situations. Finally, there is emotional learning and memory facilitated by the amygdala for the next time. This sensing, reacting, processing, and adaptation occurs, in full, three times in the film: smell, taste, and hearing. It is in adaption that we see the best of the human spirit and some excellent science shine through the apocalypse we witness unfold.

As Susan mentions in the film, taste and smell are two of the chemical senses. This categorization includes senses that detect the chemicals directly. It is like a lock and key, where the chemical is the key cells in the nose and mouth are the lock. The chemical key binds and unlocks the cells in order to send signals to the brain. This is different from hearing, for example, which senses the vibrations of sound waves.

With each loss of a sense we see the science of adaptation to a changing environment, and smartly, it is through something familiar and part of our everyday lives, food. Therefore, Michael's character is our conduit to understanding how people's lives go on. With the loss of smell, Michael increases the intensity of the remaining characteristics of the food that can be sensed, including stronger taste. Taste is the pure sensation of chemicals that are sweet, sour, salty, bitter, umami (and likely fat, the 6th taste). When this too is lost, the chefs begin to rely very heavily on the third chemical sense called chemesthesis, which includes the ability for chemicals (like capsaicin in hot peppers) to cause a pain reaction. Chemesthesis is part of the sense of touch (think food texture), a sense that the characters do not lose. So this makes for a great way to keep relevant in the changing culinary scene. The chefs play with different textures, temperature, and spicy ingredients. In addition, you witness people paying attention to sound, and we see the food gets ever more beautiful and aesthetically pleasing. The need to still feel and see and taste joy in the world is the attempt to get back to normal, which includes pleasure, and avoid any conclusion that would lead to despair, to living only for fat and flour.

Resiliency of the film's population is nearly decimated when people begin to lose the ability to hear. This third stage in the epidemic comes on gradually and via the emotional volcano of anger. Combined, these factors have managed to elicit panic in ways we had not previously seen. Us and them, isolation and quarantine. Restaurants are appropriated by the government to make mass meals to sustain the afflicted public. Public servants in full hazmat suits deliver the meals and attempt to bring order to the devastated city. In irony, a dish of fat and flour (plain pasta) is delivered to his home when Michael is quarantined after losing his hearing. Despite the despair and fear, the film progresses once more. People eat, they work, and they communicate. Life adapts once more. We see now that the food is even more beautiful than before. People learn sign language. They attend music clubs where the sense of touch allows them to experience the music though vibrations. The loss of three senses is not a death sentence. It is not the end of the world. There is hope, but people inevitably prepare for what could still come.

This is the unknown: how people adapt after loss of a fourth sense, vision. It appears to occur uniformly across the world, and so there is likely little in terms of preparation on a large scale. We are forced to think about this from the framework of survival. Humans trust our eyes more than any other sense. The basic needs of food and shelter, a means of protection from harm and continuing to sustain our body, may now be more difficult than ever for people to secure, especially in metropolitan areas where one relies heavily on others. Which leads us back to connectivity.

We Make Perfect Sense

From their first moments of vulnerability and intimacy, we recognize that Michael and Susan's shared experience of succumbing to each phase of the disease, to sharing in the unknown of what it means for the world as they know it, and to adapting, leads them to connect more deeply and more genuinely than likely either thought possible. When they experience grief and the loss of smell, she receives comfort and he sleeps soundly. With the loss of taste following intense fear, they find each other and embrace as if the world is only stable when you have someone to hold you in place. The violent explosion of anger sends them apart, they experience the hurt deeply, and the loneliness of this emotion is compounded by the loss of sound. And so, when joy changes their perception of the world, they find each other once more, to face the uncertainty of survival in a changing world together. Touch, and only touch remains. "Because that is how life goes on, like that." The message then is hope, against all odds of survival.

Chapter 6: *Upstream Color*

Amy Seimetz as Kris and Shane Carruth as Jeff escape the world in *Upstream Color* (ERBP, 2013).

Upstream Color (2013) is the second feature film by writer-producer-director-actor-composer Shane Carruth. His first, *Primer*, was a small ($7000!), indie sci-fi film from 2004. That first effort was a smart, interesting, and somewhat opaque film that was both confounding and attractive to audiences and critics alike. Both films were critical successes if not box-office hits (although *Primer* found life on the Netflix DVD-order system; Remember that?). For *Upstream Color*, the same description can be used: smart, interesting, opaque, confounding, and attractive. But we can also add beautiful, sensual, mesmerizing, and mature. This is another film that at first glance might raise an eyebrow when entered into the category of science fiction. There are no aliens; no giant battles with spaceships; no space, actually; no discernible "other"; no end-of-world narrative; and it is not set in the future. I don't think, anyway.

The narrative—if we can call it that—starts with a woman, Kris (Amy Seimitz), who is abducted by The Thief (Thiago Martins) and given a psychotropic drug transmitted through a small worm in her bloodstream. The Thief mentally manipulates her, steals all her money, and leaves her broken, jobless, and confused. She is drawn, by sound, to a man (the Sampler—Andrew Sensenig) who surgically removes the worms and implants them into a pig. He delivers the pig to a large, rural pen (and we get the feeling he has done this before). Later, Kris connects with another man (Jeff, played by Shane Carruth), with whom she may have shared the same experience. They develop a relationship that leads them to find out more about that experience, later finding their connection to others in similar situations, and finally, reuniting with "their" pigs.

To describe the narrative in this fashion, however, is akin to describing *Moby Dick* as a story about a guy and a whale. The structure is elliptical and somewhat confusing; yet there is method: the *mise-en-scène*, cinematography, editing, and sound all add up to a film that is best *felt* rather than understood. Carruth assumes an intelligent audience, allowing us to ride the stream along with the film rather just "figure it out." This is not a film that answers all its questions or ties up everything neatly. It is a film that pushes us toward looking inward and trying to answer that essential question of science fiction: "What does it mean to be human?" The film shows us that human nature pushes us forward by telling stories, listening to stories, and trying to make some sense of them. If we can't make sense, it may be either confounding or liberating. Either way, the film leads in a (perhaps not so clear, but wonderfully poetic,) specific direction: connection. To be human is to crave, seek, and understand real connection, says Carruth through his film. The connection between humans in this film is the obvious path, but even more appetizing is that the connection goes one step further: a connection to the animals and the physical world. It is a beautiful and touching connection on many levels. In the end, it goes directly to the heart of science fiction.

The film may be difficult to follow and one in which we can find a clear narrative structure upon first viewing, but again, it is there. The film is thus best understood allegorically, and there is no one simple answer. The great Italian director Federico Fellini once said that his job was to bring viewers to the train station at the end of a film. Once there, it was up to the viewer to board the train of their choosing. *Upstream Color* gives us many trains to board. The beauty is not only boarding one of them but thinking through the abstractions of each journey. We happily take that journey upstream.

To begin to make sense of the narrative and to extrapolate meaning, we can start with the structure and filmic aspects. Each aspect feeds into the larger whole. Upon several viewings, it becomes quite clear that this is a fantastic formalist film—one where the *mise-en-scène*, cinematography, editing, and sound point toward a specific point. All aspects seem individually disconnected, disjointed, atonal, incongruous, and edgy. The editing is elliptical and jumpy. The cinematography is packed with close-ups and soft focus that only keep the subject in focus while the rest of the frame is blurred. The sound doesn't always match up to the images and is frequently discordant. And the *mise-en-scène* doesn't always ground us, but rather, it disorients us. The sum total points to the overall experience and the feelings the characters have as they move through the world. *In other words, the film makes us feel as they feel.* As we take each aspect, one-by-one, we can see how the parts fit and how that progression toward a coherent whole actually works.

Starting with the *mise-en-scène*, we are immediately thrown off balance. Strange plants. Dirt. Worms. We are only given bits and pieces and the focus is blurry, with only fragments of the foreground available to us while the surroundings and the background are blurry. Faces. Hands. Worms placed into capsules. We see a bit more as The Thief attempts to peddle his worm-filled horror outside a bar. Unrequited, he retires to his car and retrieves a taser. He is in a bar now, and we see Kris enter a bathroom. Dragging Kris out the back door. Legs, shoes, rain, a parking lot. This fragmented *mise-en-scène* continues like this for much of the film, and when coupled with other cinematic aspects, it will eventually start to make sense.

Cinematography helps our cause. For the majority of the film, we are given a soft focus—achieved with a telephoto lens—that only keeps one part of the frame in focus, either the foreground or background. In this case, it is the foreground, and our focus is usually on a particular subject or object—a face, a worm, a water pitcher, a book. Zeroing in on the faces for a second, the close-up is the shot of choice here. The close-up is a psychological shot, bringing us physically closer to the subject and either illuminating something about their psyche or allowing us to study something we don't quite understand; in this case, in this particular film, *it is the characters who do not understand.* The close-up also observes the character in their own space, without regard to the surroundings. Consider the complementary shot, the long shot, wherein we see the character in relation to their surroundings. The long shot is thus more of a sociological composition, and we get to observe the subject in a "relationship." We get no such shots in most of *Upstream Color*. Mostly, the characters exist in a universe of their own, as if they, alone, are left to figure it out. God's lonely person. The cinematography captures their disoriented, confusing, and solitary existence.

The editing may be the most expressive movement of this symphony. We cut from shot to shot, scene to scene, sequence to sequence, without an abundance of continuity. When The Thief holds Kris captive, the method is the jump cut—a discontinuous transition between shots. The water pitcher to the sun shining, the book to the paper chain, the kitchen table to the car. The jumpiness of the editing speaks to the character's state of mind as much as it does our own frame of reference. We are left to fill in the ellipses and make connections. That project continues after the sequence with Kris and The Thief ends, and we see her on a train. Cut to Jeff on a train as well. We know Jeff because we were given jump cuts of him from earlier in the film (without any explanation or context other than he is a guy, running). The editing that gives us separate shots of them upon their first encounter tells the story as they meet: this is two distinct individuals. They are both alone. Even as they begin talking to each other, we always see them separately, and sometimes, the sound doesn't follow along. We may hear them talking while the image cuts to them sitting alone, disembodied and fractured. The parts are adding up now: the editing evinces the same feelings of being alone, disoriented, and confused.

Sound rounds out that feeling and punctuates it. Carruth wrote the score for the film as well, which has a symphonic feel and structure to it but is minimalist and simple in its construction and application. Most of the time, the music consists of simple keyboards that evoke a feeling or emotion. Rarely does the music serve as a leitmotif or tag a character/event. Following the same project as the other elements of the film, it can be disjunctive and discordant. But sound includes more than music on the soundtrack, and dialogue and sound effects fall into this category as well. Most films rely more heavily on the former than the latter, yet this film reverses the polarity. There is so little dialogue that the finished project is more a silent film than a talkie. It is hardly talk-y. The sound effects factor in more prominently, as my colleague Dr. Nicole Garneau discusses in the next portion of this chapter. The sound effects also serve as a point of connection for Kris and Jeff. The sounds they hear begin to percolate, and they realize that they have a shared history with the sounds. They investigate the origins of the sounds, find resonance in nature and particular places, find the farm where The Sampler lives and records, buy his recordings, and finally, they put it all together and locate him. It is through the shared connection to sound that they are able to put all their pieces together. But there is more.

There is a specific cycle involved in the "scheme" shown in the film that we can read in a number of ways. The cycle begins with the women (the credits call them "Orchid Mother and Orchid Daughter") picking up the blue orchids from the stream, which they put into pots of soil. (Even though this is shown last, I see it as the first step in the cycle.) The Thief then buys the orchids and harvests the worms inside the soil. He "injects" the worms into his victims. The worms hold a parasite that exerts a psychotropic drug into the host, rendering the host subservient to The Thief. He then steals all their money and ruins their lives. The worms are now transferred from the victims by The Sampler, who takes them from the victim and inserts them into pigs. This seems to create a connection between victim and pig. He uses sound to gather the worms through a process called "worm charming" that brings worms to the surface of wherever they are (since they are attracted to sound). The victims are then attracted to the sound as well (through the worms) and that is how they find The Sampler. The Sampler then kills the pigs and leaves them in the stream, whereby the parasite seeps into the orchids and turns them blue, beginning the cycle again.

The subjects of the film are many, varied, and some, frustratingly, remain upstream and out of our reach. Rape, assault, predation, cycles, addiction, animal rights, existentialism, and most prominently, connection, are all indirectly injected into the stream of the film. The film is also a love story—as unconventional in its presentation as its construction. In his review of the film, Mark Olsen of the *Los Angeles Times* notes, "With its densely layered, thematically rich storytelling, *Upstream Color* is in part about the mutual psychosis that can be an essential part of romance, the agreement of a shared madness." We see evidence of this dynamic throughout the film, and it is no doubt there, but as Olsen notes, it is only part of the whole.

The other subjects also fill in missing pieces and speak to that whole. There are so many wonderful themes we can attach to these subjects as well. One that Carruth himself mentioned was "breaking a cycle." He was thinking about how we may extract ourselves from a particular cycle, such as violence, or addiction, or even an unhealthy relationship. We need to face it and break through, to kill it. The two most important scenes in the film, for me, are when Kris finally looks directly toward The Sampler at the table. Before that, she couldn't "see" him sitting right next to her. But now she actually *sees* him. She is facing her addiction/fear/whatever it is you want to apply. Then, when Kris actually "kills" The Sampler, she is breaking the cycle. That break becomes the key to unlock the final portion of the film, where Kris and Jeff find the boxes of artifacts, uncover the entire scheme, and then send the copies of *Walden* to the other afflicted. What are the "afflictions?" Perhaps everyone has their own, and that is one of the elements the film leaves out of reach; yet at the same time, by leaving it narratively unreachable, we are forced to look inward and uncover our own afflictions. We need to kill our own Sampler.

The connection to the natural world, and the animals as part of that world, is another prominent theme with abundant evidence layered throughout the film. The victims are literally connected to the pigs when The Sampler removes the worms and places them into the pigs. The connection goes beyond after that, however, and it becomes metaphysical. After Kris and Jeff come together, their pigs come together. They mate, and Kris' pig becomes pregnant. Kris thinks she feels this in her own body, but a doctor visit reveals that she cannot become pregnant. The editing makes the connection clear in a cross-cutting sequence that parallels Kris and Jeff with the pigs. They can feel what each other is feeling. That feeling becomes heartbreakingly real once the piglets are born and The Sampler kills them (by horrifyingly drowning them in the river). Kris and Jeff have strong, visceral reactions to this and experience anger, loss, and grief; Jeff initiates a fight with co-workers and Kris frantically searches her workplace, looking for something she lost. Once again, following these sequences, they come together (almost inexplicably drawn to each other, actually) and attempt to locate their source of grief. This connection to animals is an important subtext to the film. Specifically in reference to the pigs, Carruth himself notes in the interview with the *Los Angeles Times*:

> It's more about what those pigs are now embodying. I mean, there is a break of the cycle. These people that have been affected by this are now taking back ownership of the thing that they're connected to.... I don't believe that narrative works when it's trying to teach a lesson or speak a factual truth. What it's good for is, an exploration of something that's commonplace and universal—maybe that's where the truth comes from.

This universal truth of which Carruth speaks indicates a certain Aristotelian bent—not a universal in terms of form, as Plato argued, but rather something predicated on an existing thing, and in this case, that "thing" seems to be very much an individual experience. The truth we understand, and that which the characters find, is both tangible and existential. They are able to share that truth, make some meaning in their lives, and move forward.

We now circle back to the cinematography. Remember the early sequences of the film were a shallow focus with only the subject in focus and the rest of the frame blurry and unreadable. That focus changes as Kris and Jeff meet, date, and eventually marry (a fact made clear only through the *mise-en-scène*: rings). The frame opens up a bit to include both of them in the frame—a two-shot; yet we still do not see very much else. The gray and gloom, as if it were always a Villeneuve "rainy Tuesday morning," also pervade the film. That all changes near the end, however, in the ending sequence at the farm: we get long shots with a deep focus that tend toward long takes. Deep focus (achieved by using a wide-angle lens on the camera) allows us to see both foreground and background in focus and gives us a strong depth of field. In the farm sequence, we get to see the characters in relation to their surroundings. Everything is now in focus. We even get the added bonus of a sunny day with abundant natural light. The net effect is an objective reality for the viewer—we get to see the world as it is, not blurry nor subjective. For the characters, they have moved from darkness, confusion, and disorientation, to clarity.

With multiple people now in the shots, clarity of both foreground and background, no intrusive, confusing, or disjunctive editing, and a natural setting with natural light, all of the film's constituent elements speak to a sea change, and a positive change at that. We have togetherness and connection at the end of the film. There is also more empathy towards life in all its forms, as in the case of the pigs. The form of the film has changed to mirror the change in the characters and theme. It's not a just beautiful and interesting film, it's a wonderfully constructed piece of art. It is firmly grounded in Earth. The question of "What does it mean to be human?" is answered on several levels in this film, and it may be mean something different to everyone watching. Which train will you take?

—Vincent Piturro

Parasites and Defying Explanation

Nicole L. Garneau

First, a "Primer" (Pun Intended) on Parasites

To put it in my student Emma's words, "Just finishing watching ... um ... that was bizarre to say the least. But it did remind me of that flatworm parasite that infects snails, ants, and then cows. When it's in the ant it controls its mind causing the ant to cling to a blade of grass where it is more likely to get eaten by a cow." My thoughts almost exactly, except my mind jumped to the guinea worm first. But let's not get ahead of ourselves, because it doesn't get any better than parasites.

Parasites have been wreaking havoc since the dawn of time and documented since antiquity. From hieroglyphics to the medical texts of Hippocrates, humans have been trying to understand how something so small could cause so much damage. Parasitism is one of the three forms of symbiosis. You can have symbiosis where one organism benefits and the other is neutral (commensalism), where both benefit (mutualism), or one benefits and the other can cause harm (parasitism). All three of these forms come into play in *Upstream Color*, but it is the parasitism that carries the story and leaves us a bit scarred alongside the main characters. So that's where we start.

According to the textbook *Parasitology & Vector Biology* (2nd Edition), the term parasitism is defined as "relationship of two organisms of different species in which the smaller (the parasite) has the potential of harming the larger (the host) and in which the parasite relies on the host for nutrients and for a place to live." The host is also required for one more vital piece of survival, developing and reproducing. An intermediate host carries a immature parasite, while a definitive host carries the adult parasite and serves as the living boudoir (so to speak) to ensure reproduction. Finally, a vector is the knowing or unknowing noun (person, place or thing) that transmits the parasite to another host, and if alive itself may technically count as another host if some sort of development occurs.

All that said, if you want to understand director Shane Carruth's parasite and its impact on all the characters, we need follow its life cycle. This starts by identifying stages: the egg, the immature parasite, the adult, and reproduction which brings us back to the egg. From there we can figure out which of Carruth's characters are hosts and what factors help this lifecycle happen over and over again until it is broken. Keeping the idea of survival in mind, we can then understand the things that make people do what they do in the midst of this lifecycle, driving the actions all the different players take. Once we have all that we can see how well this fictional parasite, and the many people and things it interweaves during its lifecycle, lives up to the world of nonfiction.

The Players: Hosts and Vectors

I stopped counting at 8, but it took at least that many viewings of *Upstream Color* for me to scientifically make sense of the hosts and vectors in Carruth's parasitic mind trip. Let me save you the trouble and spell them out for you.

- River (Upstream): environmental factor of moving parasite from one host to another
- Orchids (Color): hard to determine if the orchid is an actual host or just a vector. If host, some form of developmental growth of the parasite (e.g., going from egg to juvenile) needs to happen, but there is no evidence of this. The only evidence we have that something happens to the plant is that when it takes up the eggs via

its roots, the flowers turn from white to blue. So likely the orchids are a vector, a means of transferring the parasitic eggs from the water to the caterpillar.
- Caterpillars: Likely eat the plant and ingest the parasitic eggs. If the caterpillar is a true host, then we expect the eggs to hatch once in the caterpillar and develop into stage 1 juveniles. Let's assume this happens and call the parasites tweens at this stage.
- Orchid florists: These gals are likely another batch of environmental factors, no different from the river; they carry the parasite to a new location necessary to fulfill its life cycle. However, their actions lead to a direct benefit, so this could be a form of mutualism with the parasite (unbeknownst to them).
- Thief: Another key vector, and another form of mutualism. The thief and the parasite benefit greatly from this relationship. As a vector he physically transfers the parasite to infect a new host (victim). He is not a host as he himself is not infected and plays no role in the parasites' development or reproduction. He has learned (evolved/adapted) as a vector through the selective pressure of his circumstances to take advantage of the neurological damages caused by the parasite on victims. He can influence and alter the victim's sensory system (and thus their survival system). They see what he wants them to see, and they feel what he wants them to feel. In this way he manipulates them to hand over their assets, leaving them broke and broken, and therefore less able to survive.
- Victim(s): Intermediate host, this is definitely parasitism. An infected caterpillar is ingested (forcibly by the Thief) and we watch as the parasite develops from a stage 1 juvenile (tween) to a stage 2 juvenile (teenager) within the human host. This is clearly the most dramatic host interaction in Carruth's visual depiction. Following ingestion, we see the symptoms and pathologies of infection in the human host.
- Sampler: a vector twice over. First vector role, he surgically transfers the teenage worm from the human (intermediate host) to the pig (definitive host). The victim drinks a milky white elixir (likely an anthelmintic drug, which stuns the worm). Then the Sampler slowly rolls the worm(s?) out from an incision on the foot from the human to the pig. Unclear if this is mutualism and if he benefits from people coming to his farm in that it helps him make and sell music? Or is this a form of commensalism where it doesn't hurt him (in fact he might think he is helping the victims), but it surely helps the parasite. Second vector role, the Sampler removes the infected piglets from the parents, bags them and coldly tosses

them in the river to die. We learn that this is not the first time he has done this, and therefore plays two very important vector roles in the life cycle of this fictional parasite.
- Female adult pig: Definitive host. Teenage parasites develop into sexually mature adult parasites, reproduce, and eggs form. It seems that male adult pigs are dead-end hosts, and female adult pigs are also dead-end hosts unless they become pregnant and can then pass on the eggs to their piglets/offspring.
- Piglets: Unable to escape drowning, the bodies of the piglets move downstream where they will aground near the bank and decay. As their bodies decompose, eggs are released into the river near to the root system of the white orchids.

Questions that remain that science can't answer: How did the Sampler stumble into both his vector roles? He is not in cahoots with the thief or the orchid florists, so is this a comment on selective pressure, survival and evolution? Also, why does he feel it's necessary to kill the piglets that are born on the farm from infected parents? And why drown them, and why leave them to decompose in the river? And more confounding, how did he reach the point where he attracts and has a mobile medical unit to surgically transfer worms from human hosts (victims) to adult pigs?

A Sketch of the Parasitic Life Cycle

The eggs are released into water and infect the orchids, turning them blue. Caterpillars eat the orchid (and parasite eggs, which develop into tweens in the caterpillar/intermediate host). Orchid farmers are a physical vector and move plants and caterpillars to a new location. Thief is a physical vector and moves caterpillars from orchids to another human (victim, intermediate host). Humans ingest the caterpillar (tween parasite which develops into a teen). A zombie-ism of sorts happens when the parasite wreaks havoc on the body and the human brain. Human Sampler is a physical vector, attracting victims via sound and then physically transferring the teen parasitic worms (surgically) from human to pig. Teen parasites mature to adults and reproduce. If this occurs in a female adult pig who then becomes pregnant, the parasitic eggs are passed onto offspring (piglets). Here again the Sampler is a vector, physically moving the pigs from the farm, to the river to drown, whereby they are carried down river and decompose. The decomposing flesh leads to the release of the parasitic eggs into the water, which brings us back to the beginning of the life cycle.

"They could be starlings. They could be cracklings."

Given the evidence, what are the possible real-life parasites and how do they compare? They could be guinea worms. They could be liver flukes. The scientific name of the guinea worm is *Dracunculus medinensis* and translates to "fiery little dragon of Medina." The worm that has a long and well-documented history in inflicting human suffering and according to the World Health Organization, it is the largest parasite affecting humans and is extremely painful as the worm parasite moves through the victim's subcutaneous tissues. The origin of the scientific name pays homage to the early physicians that diagnosed and treated the infected. Dating back to the Old Testament, it is believed that the guinea worm is the infamous "fiery serpent" from Numbers. This is from where the prominent Greek physician Aelius Galenus (129–200 AD) coined the disease as *dracontiasis*, Latin for "affliction with little dragons." Later, Arab-Persian physicians (9th century AD) called the disease Medina vein (Medina being the "City of the Prophet" in Arabic, the place where the prophet Muhammad is buried). So although the more commonly used Anglo name of "guinea worm" is used, it came many centuries later when European explorers encountered the disease on the Guinea coast of West Africa, therefore, the historic nature of the parasite is well-captured in its scientific name.

As the name "fiery dragon'" implies, the guinea worm is no walk in the park. The female worm has grown and made her way to the human victim's foot (the human in this case is the definitive host), causing a large and painful blister. She lies in wait as her eggs hatch into tween larva. Victims often soak their painful feet in water at this stage, seeking relief. This behavior is exactly what the worm needs from the human host. The water signals to the female adult worm to spew her tween larva into the water. Tiny crustaceans called copepods act as the intermediate host when they eat up the tweens and they develop into a teenaged larva. The crustaceans swim happily along in their freshwater pond, until another human host scoops up water to drink and unknowingly gulps down the infected crustaceans. The teenage larva comes along for the ride and then takes about 6 weeks to travel to the skin of the person that drank the water. After 3–4 months there are equal numbers of males and female parasites coursing through the person's skin, mating occurs, and by 6 months males are no longer present. Jump forward to about one-year post-gulping the infected water and we now have full blown adult and pregnant female worms. And that brings us back to the beginning of the life cycle where the victim soaks their aching feet in freshwater and the cycle begins anew.

There are drugs that can kill or stun worms, but for thousands of years the treatment used for relieving people of guinea worm is to slowly extract the worm by rolling it out onto a stick or a rod. This visual may be familiar to you: it is the staff of Ascelpius associated with the World Health Organization and many other medical symbols worldwide.

Evidence in favor of the guinea worm as Carruth's fictional parasite is straightforward: how it looks, how it moves in the human host, some of the hosts match up (copepod = caterpillar/intermediate host, human = adult female pig/definitive host), the milky drink is given to stun the worm prior to extraction, and the preferred method of rolling out on a wooden rod. Where Carruth's parasite deviates from the guinea worm is in the number of hosts, the details of the lifecycle and more intriguing, the neuroscience.

This is where my very astute student, Emma, comes back into play. She had written a blog post about said parasite and the ant zombie, and it turns out, she wasn't alone. Otherwise known as liver fluke or *Dicrocoelium dendriticum*, this guy has been extensively written about; from official reports from the Centers for Disease Control to an online comic called, "Why Captain Higgins is My Favorite Parasitic Flatworm." But my favorite is a chapter from the book *Faith, Madness and Spontaneous Human Combustion*, in which the author, immunologist, and science writer Dr. Gerry Callahan compares the zombie ants to his crazy uncle Henry. In his book, Callahan has just described Henry and how much Henry drove Callahan's mom up a wall, when he leads you down a picturesque journey of cattle grazing in the early morning. In the first iteration, the camera zooms in and you notice ant and snails, but they are low in the grass and avoid being eaten. In the second telling we see the same cattle, same grass, but the camera zooms in and we see something very different. In Callahan's words:

> Below the cattle, many of the ants have worked their way up shafts of grass and appear to be waiting for something. As we watch in cinematic magnification, a pink tongue, large as a python, wraps itself around several insect-encrusted blades of grass, brown ants, lots of them, suddenly disappear behind a huge set of cud-scarred teeth.... Fade to black. Let's consider, for a moment, what we have just witnessed. In the first scene, the ants are cautious, responsible, sane. In the second scene the ants are none of those things.... Whether we are comfortable with the words or not, the ants in the second scene are clearly crazy.... They have lost the ability to care for themselves and seemingly no longer regard life more highly than death.... Still, we don't look for histories of child abuse among ants, discuss ant toilet training, or accuse ants [like we have of Uncle Henry] of character flaws and laziness.... So if "mentally ill" isn't accurate, what should we call them...

We should call them infected. Sound familiar? The liver fluke usually finds itself in a definitive host such as a cow, sheep, or goat, but yes, humans are susceptible too. An infected animal (or human) defecates and parasite eggs come along for the ride. Snails (intermediate hosts) eat the eggs, the eggs hatch, and they develop into tween larva. The snail leaves behind a slime ball contaminated with the tween larva. The ant eats up this delicious slime ball treat, ingesting the tween larva along the way. Yum. The tween develops into a teenager once in the ant (another intermediate host). The ant becomes a zombie (a.k.a. crazy, mentally ill, etc.) and lives out its death sentence of trance-like-climbing-grass-and-waiting-to-be-eaten-by-a-cow (or human). Once in the definitive host the teenager becomes a fully mature parasite able to mate and procreate, and we have a new round of eggs that exit once more alongside feces. And the cycle starts anew.

Evidence in favor of the liver fluke as Carruth's fictional parasite is also straightforward: the alignment of the hosts and life cycle (snail= caterpillar/intermediate host, ant=human/intermediate host, cow=pig/definitive host), and the neurological change in the second intermediate host (ant/human) leading these hosts to act in ways that do not ensure their survival and go against their nature.

Love and Breaking the Parasitic Life Cycle

Parasitic diseases are controlled by a number of actions. In *Upstream Color* we see many of these actions beautifully employed by love and facilitated by death. From unconscious to conscious, Kris experiences an awakening. Her trance-like recitation, word for word, of *Walden*, captured and brought to light by Jeff, is the key to her then tapping into the Sampler's mind, his experiences, his sounds. The Sampler becomes the sampled, and this awareness leads to his death. By killing the Sampler, Kris single-handedly eliminates him as a twice-over vector. First, he can no longer attract victims and transfer their worms to adult pigs. Second, he can no longer send piglets spawned from diseased pigs to their death in the river. This then eliminates the contamination of the river and the orchids. There are no blue flowers for the florists to scout, and therefore no caterpillars for the thief to employ in stealing away the lives of more victims.

Verdict

Carruth's fictional parasite has a life cycle of the liver fluke, as well as the neurological repercussions of the fluke when in the second intermediate host. The fictional parasite, however, has the visual presentation and development of a roundworm (e.g., guinea worm) when in the human host. How this parasite brought together all these characters is the beauty in selective pressure and evolution. How it came to an end is nothing short of love, something the human mind can feel but can't really fully explain. Kind of like this film.

Chapter 7: *Contact*

Jodie Foster as Ellie Arroway listening to the galaxy in *Contact* (Warner Bros., 1997).

Robert Zemeckis may be the best and most prolific director you have never heard of. Steven Spielberg and Stanley Kubrick don't need much of an introduction, but Zemeckis may. His films, however, speak for themselves and have been part of the lexicon of American film culture as well as culture writ large: *Back to the Future I, II, & III* (1985–1990), *Who Framed Roger Rabbit* (1988), *Forrest Gump* (1994), *Cast Away* (2000), *What Lies Beneath* (2000), and *The Polar Express* (2004), among others. His films are known for their technical prowess and innovation, their quirkiness, and their top-notch production values among countless other things. Not to mention wildly successful. Many of the images from his films have been burned into the collective psyche, from Michael J. Fox in the time machine; to the animated, curvy Jessica Rabbit; to Tom Hanks eating chocolates on a bench; to Tom Hanks skinny as a stick on the beach in *Cast Away*. We can go on and on. Zemeckis' career spans six decades and counting, and his influence looms large over recent American cinema.

Zemeckis grew up in Chicago in a blue-collar Catholic family, leaving for USC film school in the late 1960s. While at USC, in the midst of the European art film invasion and the beginning of the American New Wave, he was unimpressed with that scene and was more interested in making classic Hollywood films. It is there that he met writer and future collaborator Bob Gale. After being recognized for a student film, he lobbied Steven Spielberg for help; Spielberg was impressed and produced his first two feature films *I Wanna Hold Your Hand* (1978) and *Used Cars* (1980). Neither film was a box-office success, but everyone recognized the writing (in partnership with Gale) and directing talent. Struggling for work, he made *Romancing the Stone* (1984), which everyone thought to be a dud yet became a surprise hit and allowed him to make the film he and Gale had trouble selling to all major studios: the story was of a teenager who inadvertently travels back in time to the 1950s. The rest was, well, the future.

His career entered the stratosphere after the breakout success of *Back to the Future*. In between making the sequels (something many big directors have frequently refused to do), he made the groundbreaking and technically brilliant *Who Framed Roger Rabbit*, known for its mix of live action and animation. After several other films, he made *Forrest Gump* and won the Oscar for Best Director (among many other awards for the film). With his penchant for big projects and his deft hand at technology and storytelling, he then turned his sights on the adaptation of Carl Sagan's popular—and dense—science fiction novel, *Contact*. Originally conceived as a film, Sagan changed his approach and published it as a novel in 1985. The story of contact between humans and extra-terrestrials became a best seller, sold millions of copies, and subsequently went into production at Warner Brothers, who owned the rights from its initial iteration as a film treatment. The script went through several drafts, writers, and directors, including Roland Joffé (*The Killing Fields*) and George Miller (the *Mad Max* films). Zemeckis finally took the helm and was given complete artistic control (after originally turning it down), including final cut rights—a rarity in the big studio world. With a final budget of $70 million (a gigantic figure even by current standards), A-list stars including Jodie Foster in the lead role, and a sprawling narrative, Zemeckis seemed the perfect person to make the film.

The narrative of the film centers around Dr. Ellie Arroway, (Jodie Foster, and Jena Malone as the young Ellie) a SETI (Search for Extra Terrestrial Intelligence) scientist whose work has her listening to the stars for signs of life. Her funding is eventually cut, and she finds a sympathetic patron in eccentric billionaire S.R. Hadden (John Hurt). When she finally discovers what she thinks is such evidence of alien communication, Dr. David Drumlin (Tom Skerrit), the president's science advisor, takes over and heads up the project to build the spaceship made from designs communicated from the "aliens." He is also chosen to take the ride after Ellie is deemed not fit for first contact since she does not believe in God. When a religious terrorist (Jake Busey) blows up the machine (and Drumlin) on the launchpad, Hadden reveals he had secretly built a second machine and installs Ellie as the occupant. Her ride seems to take her to another galaxy, through a wormhole, to a planet where she makes first contact with beings who take the physical shape of her long dead father (David Morse). Her "father" tells her such a shape was made to make her feel more comfortable with the meeting. The meeting is short, and Ellie is catapulted back to the capsule, where she finds her cameras did not record anything, the capsule never actually went anywhere, and nobody believes her story (despite some evidentiary proof to the contrary). As everyone shuns her and accuses her of engaging in a corporate sham (run by Hadden), her friend and sometime boyfriend Palmer Joss is there to console her in the end, tying together science and religion in a lovely bow.

With Jodie Foster installed as Ellie, Zemeckis brought in Matthew McConaughey as Joss, the theologian with whom Ellie as an on-again/off-again relationship throughout the narrative. The story is similar to the book but changed for the screen. The biggest change is that Palmer Joss is no longer a fiery, tattooed, preacher man; in the film he is a soft-spoken author who serves as Ellie's love interest and intermediary between stronger religious voices who are opposed to the journey. Other characters have been deleted (as is standard for a screen adaptation), and others have been collapsed. Of course, the book has more room for philosophical debate and discussion, and the film condenses the five characters chosen to take the ride into one, Ellie. In the book, the five characters are chosen for different reasons—including cultural and international inclusion—all adding to that discussion and debate. The film shorthands this for dramatic purposes as well as economy, but in the process loses some of the philosophical depth and makes it much more American-centric. Sagan's book, written toward the end of the Cold War, is driven, in part by the hope that finding such extraterrestrial life would bring the world together in a common cause. The film includes some of that sentiment, but it all ends up in America, with Americans. One of the more interesting changes is that in the book, when the ship returns, there is more evidence for the actual journey—including sand in the capsule. Yet the crew of five is accused of an international conspiracy (shedding doubt on any Cold War warming) and sworn to silence. For her part, Ellie finds out a secret about her real father and tries to uncover more about the trip—through math. Yeah, Hollywood didn't go for that part.

The film, however trite the ending, still manages to infuse a great deal of '90s politics, culture, philosophy, economics, and technology into its own narrative. It also has a sense of film history as well a close connection to sci-fi of the past. As some of the cast and crew noted, Zemeckis made them all watch *2001: A Space Odyssey* before starting work on the film (a curious request from someone who was admittedly more of a fan of Hollywood cinema than art cinema). Sagan also gave cast and crew lectures on astronomy, and he tried to stay involved in the production even though he was very sick at the time (he passed away just six months before the premiere of the film). The narrative steadiness as well as the technical bravado that Zemeckis had cultivated in his career up to that point served him well during the making of this film—a technical marvel that advanced CGI and green-screen technology leaps and bounds from when Spielberg used it in *Jurassic Park* just a few years earlier. The combination of the serious sci-fi source material from Sagan, along with a serious and technical director in Zemeckis, along with a serious and studied lead actor in Foster would produce a serious and technical film that appealed to many when it premiered in 1997. It would also neatly and interestingly address our central question of sci-fi: "What does it mean to be human?"

The film's opening announces its seriousness, its technical prowess, and the central premise of contact with aliens in a virtuoso sequence. The opening of Earth as seen from near-orbit with the sun peeking over is a classic sci-fi shot and hints that someone else is watching (remember the opening of *2001: A Space Odyssey* looking back at Earth), or in this case, listening. The *mise-en-scène* is beautiful and purposeful right from the beginning. We hear a panoply of audio bits from various and multiple sources: music, telephone conversations, space shuttle/Mir space station communications, and other snippets as the camera tracks back from Earth and travels through the solar system. The audio clips get older as we move further from Earth, ultimately finishing with the first audio broadcasts to make it into space in the late 1930s. As the camera finally leaves the galaxy, the audio is silent, but the camera keeps tracking back, finally stopping as it reaches distant galaxies. The view along the way is quite remarkable, especially for the casual stargazer—seeing the Moon, Mars, Saturn, and Jupiter *et al.* is thrilling and visually stunning. The *mise-en-scène* is thus infused with that sense of sci-fi history pointing back to Kubrick (being observed by an alien culture who then initiates contact), a technical brilliance, and also a sense that it seems mathematically improbable that we are alone in this vastness. As the camera finally pulls back on young Ellie and she talks to her father, he utters the line that is most often attributed to Carl Sagan and seems quite apt after that opening: when Ellie asks him if we are alone in the universe, he replies, "If it's just us, it seems an awful waste of space."

There are countless other examples of expressive *mise-en-scène* in the film apart from the opening. In that second scene of the film, young Ellie is listening to her long-range radio and picks up a communication from someone in Pensacola, Florida. Before bed that night, Ellie draws a picture of how she sees Pensacola—a beautiful beach surrounded by palm trees. That same picture will come to life later in the film when the alien version of her dad materializes on a beach that closely resembles the picture. Assuming that the aliens probed her mind to find a comfortable visage and bodily form with which to communicate, the picture was plucked from her mind as well. That drawing from childhood—a reminder of beautiful and simple times when the universe was full of potential—was burned into her and serves as a reminder to all of us how that power of youthful innocence and infinite possibility may never leave us.

Another picture looms over the film as well: a print of a Carl Sagan favorite hangs on Ellie's wall in her small apartment in Puerto Rico. The painting by John Lundberg shows what looks to be an ice planet and moons/stars around it. It pops up several times during the film: that first time in Arecibo, again in her apartment in New Mexico when she is working at the VLA (Very Large Array), and finally, when she goes through the wormhole near the end of the film, she sees the real-life version materialize outside the capsule. The ubiquitous quality of the print and its importance to Ellie signifies not only her love for the cosmos and her perseverance in communicating with other life, but it also pays homage to Sagan, his vision, his own hopefulness about life outside Earth, and how he brought all of that energy and curiosity to a whole generation with his PBS *Cosmos* documentary series (originally aired in 1980). Sagan brought astronomy into our homes and inspired us to look to the stars. Billions and billions of them. Ellie never stopped reach for them, and the painting is a continuous reminder of all those elements that make the film smart, engaging, and inspiriting.

The cinematography, like most other great sci-fi, works hand-in-hand with special effects to create the world of the diegesis while providing depth, meaning, and incredible detail. Even though it is set in the contemporaneous society (the '90s), there is still a plethora of effects that go into the film, even after the insanely difficult opening sequence. The extended sequence with Ellie in the machine and traveling to the other galaxy (perhaps? We think…) is a great example of this confluence. As Ellie approaches the island of Hokkaido where the second machine had been built, the shot is a composite of several different things that is very commonplace for the film and a great example of how it all works. The ship cuts through the water toward a bay where the mountains meet the ocean. The ship cutting through the water is real and filmed by the crew off the coast of Los Angeles. The cut to the wider view of the ship is CGI, with CGI people on the deck. As we cut to the reverse view of the ship heading toward Hokkaido, the mountains surrounding the site are also real; the crew filmed this is in

Newfoundland. The Machine site is CGI, and the sky, including the rain, is CGI as well. The entire sequence is a blend of beautiful cinematography and special effects to give the sequence the drama, darkness, and ominous feeling it finally imparts. While there a great number of effects and CGI in the sequence, it is still based in reality. Each part of the film has its own purpose, and Zemeckis is able to deftly conduct this intricate symphony.

The combination of all the cinematic aspects also come together wonderfully in the film to tell the story and elucidate its themes. A famous Zemeckis shot is the long take (an uninterrupted shot without any editing) while the camera is moving—sometimes in interesting directions, such as in a circle, or sometimes simply tracking toward or away from the subject. The technical aspects of such shots are supremely difficult, since the lighting and lenses have to be properly calculated for the entire shot as the camera is moving around. The backgrounds also have to stay in focus and keep their intensity as well, often with several different levels of illumination in that background (a sky with stars, for example). The *mise-en-scène* also plays a part in this, as all of the subjects have to move perfectly in and out of the frame at the correct times. This perfectly choreographed dance is a ballet that is performed in the middle of a modern dance surrounded by an opera. One example of this type of shot—and its wonderful expressiveness—comes early in the film as Ellie meets Joss in Puerto Rico and they, along with and the rest of the Arecibo (and later LVA) crew, including Drumlin, are having drinks at a restaurant. The camera tracks in slowly and then moves around the group as they talk to each other. Locals and others mill and talk in the background. It is night and the stars glimmer in the background. The shot seems simple enough, but it is nothing but. The technical difficulties are many, including maintaining lighting and focus levels throughout the shot as the camera moves. In many films, and with many other directors at the helm, this may have been achieved by a static camera and a few cuts of editing: a medium shot of everyone talking, and then cuts to close-ups of each individual as they take their turn. The lighting and framing are much easier in this manner, and that is standard practice on most films. But this is Zemeckis, and while technically complex, it also has thematic importance—on several levels. First it gives a sense of realism as the audience is allowed to sit and observe the shot as it is happening in real time. It also allows us to observe the characters in relation to one another; for example, Drumlin is standing in the middle, holding court, as everyone gathers around him. He is obviously *the guy*, the center of attention, and clearly loving it. Ellie is standing just off to the right of frame, not basking in Drumlin's glow, thereby placing her as not necessarily a part of his inner circle and at odds with him. This dynamic will play out during the course of the film in many ways. The first cut we get during this sequence is to Joss, who inserts himself into the conversation and adds in his theological perspective. The cut to him standing alone sets him apart as well, and even more so in the case of Ellie. In the long take of the group, Ellie is *a part of the group*, but the cut to Joss sets him apart. And as the camera cuts again, the framing is particularly expressive: Joss stands by himself to the left, Drumlin and his group stand in the middle, and Ellie is off to the right. The philosophical and ideological dynamics are set here: Ellie takes the purely scientific view of everything unfettered by politics; Drumlin takes a scientific view but it is pragmatic and always tempered by politics; and Joss takes the theological view. The *mise-en-scène*, cinematography, and editing all work together to tell us about the characters, their standing as part of the group, their relationships to each other, and finally, the thematic stakes of the film. The final cinematic element, the sound, as you will see in the second part of this chapter, has its own function as well.

Ultimately, *Contact* is a classic Hollywood film that tells its story in a smart, interesting, and technologically proficient manner that accomplishes its themes in many ways. The Zemeckis method of combining real-life elements with special effects and now, CGI, all come together to tell a great story and deliver a professional and impeccable production—just like he did in his more famous films, such as *Back to the Future, Who Framed Roger Rabbit?*, and *Forrest Gump*. He also succeeds in delivering grown-up sci-fi in the tradition of *2001: A Space Odyssey* before him as well as *Interstellar* and *Arrival* following him. The strong lead female character of Ellie was also ahead of its time and was a progenitor for such important roles such as Dr. Louise Banks in *Arrival* or Dr. Murphy Cooper or Dr. Amelia Brand in *Interstellar*, while obviously inspired by the groundbreaking role of Ripley (Sigourney Weaver) in *Alien*. The film is grounded in Earth, and it also addresses many of the classic issues of sci-fi, especially going back to *2001: A Space Odyssey* and forward from there; the central question of sci-fi, "What does it mean to be human?" has a few different answers, but the predominant theme seems to be that to be human is to keep looking to the stars, keep hoping for better, and to keep questioning our place in the universe while fighting for unity on our own planet. These answers point directly back to Sagan, but Zemeckis and the film make them their own. There are many other subjects along the way with their attached themes—the co-existence of science and religion, for example—but the film's central argument implores us to go beyond those binaries and keep looking out into the universe for meaning. That view helps us understand our own place, and therefore, the search for meaning is the biggest takeaway. And really, isn't that what makes us human?

—Vincent Piturro

The Sounds of *Contact*

Naomi Pequette

How would you react if you found out we aren't alone in the universe? Imagine the moment you discover a radio signal from another civilization had traveled billions of miles through interstellar space, had been detected by some of the most powerful radio telescopes in the world, and decoded by scientists. Would it matter if it was first detected by scientists from your home country? Would the content of the signal matter? Would you want the chance to be able to meet the alien civilization that sent the signal? These are all questions that the movie *Contact* explores.

The opening sequence of *Contact* sets the scientific basis for the rest of the film. As the camera travels away from Earth, the audience hears a cacophony of sounds. These sounds, which are radio and television signals traveling out into space, get older and older as we zoom past planets and asteroids. Eventually there is silence as the audience is taken into deep space and past beautiful sights like the Eagle Nebula. While the premise of the sequence has its basis in science, the scale is completely wrong.

Humanity has been transmitting television and radio signals into deep space for over a hundred years. These signals leave Earth and travel at the speed of light. This means that in one year, a signal will travel one light year into space. This has created what scientists call the "radio bubble," an ever-expanding sphere with Earth at the center, that spans over 200 light years and announces humanity's presence to the cosmos. These signals have gone well beyond our solar system and out to the nearest stars. However, our own solar system is small in comparison to this vast bubble since it spans just a few light hours across. That means, when *Contact* was released in 1997, our solar system would have still been listening to the greatest hits of 1997, like the number one Billboard song "I'll Be Missing You" by Puff Daddy and Faith Evans, not broadcasts of the Kennedy assassination like we hear at Jupiter during the opening sequence. The closest star, Proxima Centauri, is only four light years away, which means any aliens on the planets orbiting Proxima Centauri would be singing along with Whitney Houston's "I Will Always Love You." The television signal featuring Hitler at the 1936 Olympic games would have been traveling through space for 61 years, meaning any planet within 30 light years from Earth could have received the signal and sent it back to Earth. This includes more than 20 planets discovered as of 2019 and the all-important star of the film, Vega.

The story in *Contact* closely parallels the story of the Search for Extraterrestrial Intelligence Institute (SETI). One of SETI's first projects, Project Phoenix, used radio telescopes to search for narrow-band radio signals, or signals that are at only one spot on the radio dial. These are considered the "signature" of an "intelligent" radio transmission. Much like Dr. Arroway's research, Project Phoenix heavily relied on existing radio telescopes, such Arecibo. Despite this, Project Phoenix was still the world's most sensitive and comprehensive search for extraterrestrial intelligence. Unfortunately, this dependence on existing equipment meant that there were multiple projects competing for observing time. Still, SETI was able to obtain two three-week observing sessions on Arecibo, the world's largest radio telescope, each year between 1998 and 2005. Instead of broadly scanning the sky, Project Phoenix targeted Sun-like stars within 200 light years since they were believed to be the most likely starts to have a planet capable of supporting life, and thus possibly intelligent life. Nearly two billion channels were examined for each target star.

SETI faced funding woes much like Dr. Arroway. Less than a year after founding the program, NASA withdrew funds from SETI due to pressures. While there was, and still are, questions about whether we could find evidence of extraterrestrial life, most informed parties agreed that SETI was pursing worthwhile and valid science. However, fervor to decrease the federal deficit and a lack of support from other scientists and aerospace contractors made it an easy program to cut. Since then, SETI has been dependent on foundations and private donors for funding.

We see this reflected in *Contact* in Dr. David Drumlin who often questions the value and chance of the success of Dr. Arroway's search. Dr. Drumlin is science equivalent of a mustache-twirling villain. He will tell politicians whatever they want to hear, is narrow minded with the power to make or break scientist's careers with funding, and is the stereotypical patronizing "mansplainer" that makes him reprehensible to the audience, or at least to an audience of scientists. He represents the politicians and other scientists who often mocked SETI. "What's wrong with science being practical, or even profitable?" he muses. There is no immediate return on a search for extraterrestrials and that is often the factor that determines what projects receive funding. This was especially true for national funding of science in the 1990s. During Dr. Drumlin's visit we hear other scientists at Arecibo scrambling to justify their own research in hopes that they can keep their funding. Dr. Drumlin ultimately pulls the plug on Dr. Arroway's funding from the National Science Foundation, forcing her to seek funding from private sources. Her research became dependent on funding from private donor, S.R. Haden, much like SETI's research.

SETI served as the inspiration for key scientists as well. Dr. Arroway was based on Dr. Jill Tarter, the former director of SETI and the person responsible for the fact that SETI even exists. Like Dr. Arroway, she was inspired and encouraged by her father to pursue engineering and science before he died when she was twelve. She had to elbow her way through school at a time when women didn't pursue STEM careers and was often not respected by peers because searching for extraterrestrial intelligence was, and still can be, considered fringe. However, like Dr. Arroway, Dr. Tarter persisted and left behind an incredible legacy. Dr. Kent Clarke was based on Dr. Kent Cullers, a project manager for SETI. Dr. Cullers was the first blind student to earn a Ph.D. in physics in the United States and is believed to be the first astronomer who was blind from birth. He developed and implemented complex computer algorithms to sift through mountains of radio signals and search for one that might be from another civilization.

One key difference between Dr. Arroway and Dr. Clarke's search in *Contact* and SETI is the telescopes they used. While both Dr. Arroway and SETI utilized Arecibo, SETI never used the Very Large Array in their search. Not only would this have been a significant drop in sensitivity (Arecibo has four-times the collecting area, so it would be more likely to be able to detect a faint signal), it would have created a logistical problem. Since the Very Large Array is made up of 27 radio dishes, this would have required 27 specially designed receivers (one for each telescope) which would have been impossible with SETI's limited budget.

And forget trying to listen to all those radio signals. While Dr. Arroway sitting in the desert listening for a signal is one of the most iconic visuals of the film (and one visitors of the Very Large Array love to re-create) astronomers don't listen to signals at all. During Project Phoenix, using only one radio dish, there were 28 million radio channels being monitored simultaneously. Headphones could only listen to one of these channels at a time so the chances of listening to the right channel when the signal arrives is "astronomically" small. Unfortunately, the life of a radio astronomer is not nearly as romantic. It involves a lot of sitting in a control room (with no Wi-Fi or cell phones since that could produce a signal radio telescopes could pick up) waiting for a computer (using complex programs, like those developed by SETI's Dr. Cullers) to send an alert that there is an interesting signal. However, astronomers are required to make critical decisions about signals that look intriguing.

Much like radio signals we have broadcasted into space in hopes of contacting an alien civilization, scientists speculate that any signal we receive from an intelligent civilization would be distinct from other naturally occurring radio sources. This could be done with the content of the message, like the "Arecibo Message" sent in the 1970s which contained the numbers one through ten and information about our DNA. Certainly, prime numbers or information on how to build an advanced machine would qualify the signal in *Contact* and make it distinct. In reality, however, it could take years to decode the deeper signals so there needs to be something else to make scientists look twice at a signal.

The aliens in *Contact* do this by transmitting the signal at a very special frequency that wouldn't occur naturally. This frequency, 4.4623 GHz is described as "hydrogen times pi (π)." The hydrogen line, which is a common observation in radio astronomy, is the frequency at which hydrogen atoms, the most abundant substance in space, emit radio waves (1420.40575 MHz). While there aren't a lot of loose hydrogen atoms in space (about one per cubic centimeter of interstellar space) space is vast. So, the collection of all those individual atoms makes for a powerful signal that can be easily detected by small radio telescopes. By multiplying this frequency, that would be well known by scientists, by a mathematical constant, not only are they creating a signal that could not be naturally occurring (since pi is an irrational number), bit would also give the civilization on the receiving end clues to the scientific knowledge of the aliens that sent it. While this frequency isn't inside the range of frequencies that was observed by SETI's Project Phoenix, it is within Very Large Array's observing range of 1–50 GHz.

Another clue that the signal in *Contact* was not likely to be one that was not from a typical astronomical source is its strength. The signal measured in at 100 Jansky (Jy). A Jansky is a unit used by radio astronomers to describe the "brightness" or strength of a signal. Celestial radio sources are much fainter than terrestrial and are just a few Jy in strength. So, this is a relatively strong signal. The Sun, the brightest celestial radio source is 106–108 Jy in most frequencies, depending on solar activity. Terrestrial radio broadcasts, such as those we listen to on FM radio can be a million to a trillion times brighter than the Sun. So, while strong by astronomical standards, this is still a very faint signal by terrestrial standards and would require a radio telescope to detect.

So, what would happen if a signal is detected? In *Contact*, we see mixed reactions—excitement, wonder, fear, a sense of loss of control. The closest we have gotten as a society was on October 30, 1938, when CBS Radio systems broadcasted a story that Martians were attacking Earth, starting with a small town in New Jersey. While reports are mixed on whether there was nationwide panic or people simply enjoyed the broadcast of "The War of the Worlds," many scientists have used this reaction to frame their recommendations for "first contact" protocols. Today's society is used to getting constant updates via Twitter and other social media, so the post-detection protocols, which were first written in 1989, were revised in 2010, and are currently undergoing another revision.

As in *Contact*, the first step would be verify the signal. Since 1997, scientists have become even more connected globally which fosters collaboration and allows for this sort of testing. In an ideal situation, only after the signal had been verified would the world be alerted to the discovery via a press conference. However, in this increasingly connected world with more "news leaks" this is unlikely to happen. The 2010 International Academy of Astronautics (IAA) post-detection protocol, which is only 2 pages long, now includes informing the public earlier in the process than the original version. If the public were to find out before the signal was fully verified, scientists would manage the public's expectations by using the Rio Scale, a scale which indicates how likely the signal is to be from an intelligent extraterrestrial civilization.

Could a discovery of this possibly be contained by one government like the United States attempts to do in *Contact*? If the signal is discovered by SETI, which is not funded or controlled by a U.S. governmental agency, it is unlikely. Step three in the IAA post-detection protocol is "[a]fter concluding that the discovery appears to be credible evidence of extraterrestrial intelligence, and after informing other parties [researchers or organizations involved in the detection] to this declaration, the discoverer should inform observers throughout the world through the Central Bureau for Astronomical Telegrams of the International Astronomical Union, and should inform the Secretary General of the United Nations in accordance with Article XI of the Treaty on Principles Governing the Activities of States in the Exploration and Use of Outer Space, Including the Moon and Other Bodies." Yes, astronomers send out "telegrams." However, today they are digital and are used for all major astronomical discoveries that need further observation. This is widely used for the discovery of new supernovae which are some of the brightest phenomena in the universe but fade very quickly and need quick reactions from observatories around the world to maximize observation time. This step in the IAA protocol also includes notifying eight other international organizations. Step five requires the release of all data necessary to confirm detection to be released to the international scientific community. Unfortunately, there have been no confirmed signals yet and there are Dr. Drumlins in the world who would work closely with politicians so, despite the international community's best effort, we won't know until it happens.

So how would we react as a society? Michael Varnum of Arizona State University investigated just this. In his study, published in the *Frontiers of Psychology* in 2018, he found that we might react better than science fiction might lead us to believe. Varnum and his team ran several relevant new stories through a language-analysis program and asked it to determine whether the language used in those articles reflected positive or negative emotions. These news articles included stories about the 1967 discovery of pulsars whose regular, repeating signal was first labeled "LGM" for little green men, stories about the "Wow!" signal from 1977 which is the most likely candidate for an extraterrestrial signal but has never been verified, the 1996 "discovery" of fossilized microbes in a Martian meteorite, and more recently articles about the discovery of earth-like exoplanets and the strange behavior of Tabby's star, which was thought by some to be acting like an "alien megastructure." These articles generally turned out to include language reflecting more positive attitudes. The second phase of his study was to conduct surveys of approximately 500 people on their anticipated reaction if we discovered (and verified the existence of) microbial life along with asking another 500 people to read, and write down their reactions to, articles about the 1996 "discovery" of microbial life (now known to be incorrect) as well as an article about the creation of synthetic life here on Earth. In both cases, participants used more positive than negative language. However, this study has been criticized for its focus on microbial life. After all, as SETI scientist Seth Shostak points out, microbes are one thing and little grey aliens with an advanced technological society are another. The reality will be much more complicated than people reading a single article and writing down their reactions. People will be influenced by not only how the story is presented, but also by reactions on social media and their friends. This study also didn't investigate the effect religion will have on people's reactions, a central theme in *Contact*.

If a signal from an intelligent alien civilization is ever detected, it will be a world-changing, paradigm-shifting event. So what are the chances there is life out there that could send such a signal? "There are 400 billion stars out there, just in our galaxy alone. If just one out of a million of those had planets, and just one in a million of those had life, and just one out of a million of those had intelligent life, there would be literally millions of civilizations out there." Dr. Arroway's numbers aren't quite correct and are pessimistic even by the lowest estimates by astronomers. However, even with those numbers, it's clear that if there wasn't intelligent life out in the universe, it would be an awful waste of space.

Chapter 8: *Jurassic Park*

Sam Neil as Dr. Alan Grant attempts to outrun the dinosaurs in *Jurassic Park* (Universal Pictures, 1997).

Steven Spielberg has been one of the most prolific, popular, and successful directors in the history of cinema. His career has been long and fruitful, dating back to his days as a student at Cal State Long Beach (he was denied entry into USC film school) in the late '60s. As an intern for Universal Studios, he made a short film called "Amblin'" and was subsequently offered a directing contract by the studio (at the age of 22!). He dropped out of school and started making TV shows and films for Universal. After two features, he was offered the chance to direct *Jaws* (1975), based on the popular book by Peter Benchley. With a sizable budget and a notable cast, the production encountered numerous difficulties and cost overruns and almost didn't make it to the screen (Spielberg even feared for his career at that point). But make it it did, becoming the first American film to gross over $100 million (overtaking *The Godfather* as the highest grossing film of all time) on its way to making several hundred million dollars worldwide. The film also changed the way we go to the movies, moving the blockbuster season from the typical Christmas week release to the summer. Because of the film's beach/summer setting and the growing popularity of air-conditioned malls with megaplex theaters attached, the film was slated for release in June 1975. It was a sensational hit and played in theaters for months. The wild success off the film kicked of the era of the summer blockbuster, followed by *Star Wars* in 1977 (the film that would finally eclipse its box-office record). Spielberg himself became a multimillionaire and earned the right to make his own way in Hollywood, turning down *Jaws* sequels and only making the films he wanted to make. By the time he came to *Jurassic Park* in 1992, he had directed *Jaws*, *Close Encounters of the Third Kind* (1977), *Raiders of the Lost Ark* (1981), *E.T.* (1982), *Indiana Jones and the Temple of Doom* (1984), *The Color Purple* (1985), and *Indiana Jones and the Last Crusade* (1989), among others. Steven Spielberg seemed the perfect filmmaker to direct a big budget "monster movie" about dinosaurs engineered back to life. Not even a hurricane could stop him, and in the process, he would change filmmaking forever.

Throughout the course of film history, a combination of models and stop-motion animation were used for creating special effects, going back to the groundbreaking production of *King Kong*. During pre-production, this was the plan for *Jurassic Park*—they would use models for the close-ups and stop-motion animation for the wide shots and shots of the dinosaurs running. Famed special-effects artist Stan Winston was charged with making the models for the film, and he was well into pre-production when Industrial Light and Magic (ILM—the special effects company originally formed by George Lucas on the *Star Wars* films) approached Spielberg to do some CGI on the film. CGI had been used sparingly in film up to that point, usually for only a shot or two, but ILM was tasked with doing whole sequences. When Spielberg saw the rough cuts of the dinosaurs running across a field, he turned to effects supervisor Phil Tippet and said "It looks like you are out of a job," to which Tippett replied, "You mean extinct?" The line would make it into the film, and Spielberg decided to go with CGI dinosaurs for most shots in the final cut, excluding the close-ups. ILM would work with the animators and together they would create the sum total of effects for the film, but the era of CGI was born, changing the course of filmmaking forever.

Interestingly enough, the physical elements would also have a say in the film. Originally, Spielberg wished to film on location in Cost Rica, but the lack of infrastructure was too much of a problem, and the production team opted for Kauai, Hawaii, instead. They were able to construct the needed buildings and seamlessly blend them into the natural environment, and the striking coastline and rolling hills with beautiful vistas added to the mystery and romance of the narrative. But nature had other ideas, and Hurricane Iniki rolled through in the middle of the production, making landfall exactly where the team's hotel was located. The production had to shut down for several days, and rather than have everyone sit idly around, the cast and crew were put to work on the local cleanup effort after the Hurricane devastated the island. Never one to miss an opportunity, Spielberg had his camera crew film footage of Iniki as the storm raged, and that footage would make its way into the film. The combination of nature and technology conspired yet again to show up in the finished film.

The narrative of the film stuck fairly close to Michael Crichton's bestselling book, on which the film was based. But there are notable changes, both in the science and in the characterizations. The premise of book and film are the same: an enterprising and entrepreneurial billionaire genetically engineers dinosaurs back to life and opens a theme park on an isolated island. Things go downhill from there. Most of the characters from the book remain in the film, but they are changed: Dr. Hammond (Richard Attenborough) is much more venal in the book and instead, loving and avuncular in the film; Dr. Grant actually likes kids in the book; the "love triangle" in the film between Dr. Grant (Sam Neill), Dr. Sattler (Laura Dern), and Dr. Malcolm (Jeff Goldblum) is not in the book; and the kids are reversed in the film (the boy is the older, computer nerd and the girl is the younger one, obsessed with dinosaurs). Crichton was a consultant on the film, and he approved of the changes, which all seemed pretty common for an adaptation. His core narrative remained intact, and most importantly, the themes of the book are still present in the film. In Spielberg's hands, those themes shine through in all aspects of the film, starting with the *mise-en-scène*, cinematography, editing, and sound.

The *mise-en-scène* may be the most stunning aspect of the film, including the settings and subjects in particular. As noted, Spielberg wished to stay true to the book and film in Costa Rica, but the lack of infrastructure and the need to build the roads and necessary structures would have been too intrusive on the local ecosystem. The careful thought about disrupting the natural world given by the production stands in stark contrast to the one of the main themes of the film: disrupting the natural order through genetic engineering. Due to the smaller footprint the film would make in Kauai, the production moved there. While the wild and natural beauty of Cost Rica would have been equally stunning, the combination of lushness and the feeling of remoteness in Kauai would serve the production well and make for stunning vistas. The sequence of the helicopter approaching the island and its landing in a small valley give us the sense that the island is both beautiful and treacherous. Nature can be inviting as well as brutal. And of course, the hurricane that interrupted the production was devastating to the island but helped to make the final cut of the film more authentic. Nature gives and takes.

The *mise-en-scène* also encompasses the subjects, and the dinosaurs—both the models and the CGI-renderings—are awesome, inspiring, terrifying, and the obvious stars of the show. The sequence at the archaeological dig sets us up for the story/visage of the Velociraptor, but it is nothing to compared to actually seeing it live and in attack mode. The opening sequence of the Park worker being mauled by the Raptor is equally gripping and horrifying. Holding off until later in the film to show us the Raptor is a stroke of Spielberg genius; we get just enough information to invest the later sequences with the proper gravity and scariness. In the meantime, we are also given beautiful and majestic sequences of dinosaurs that allow us to discover their awesome realization along with the experts brought into the Park to examine.

The initial scene when Sattler, Grant, and Malcolm see the dinosaurs running across the field is a revelation for both them and us. As they discover the dinosaurs, we, as viewers, discover them as well, and in the process, unlock this new CGI-rendering that has now become so commonplace. That newness was particularly tricky for the actors on location as it was filmed; they were the first actors who were really required to react to *nothing*. Laura Dern recalls being given a sharpie "X" on a piece of paper raised up on a crane to which she needed to react as she "saw" a Brachiosaurus. She and actors Sam Neill and Jeff Goldblum were tasked with the oddity of reacting to an idea, rather than a specific object in the setting. Of course, it is not exactly the first time in film history this had happened, and in the *Star Wars* films in particular (along with other '80s sci-fi films) the actors were reacting to blue screens while the special effects were added in later. Watching a black screen where a planet would blow up in the background is quite different, however, than reacting to a *dinosaur* that will be digitally placed in the same shot next to the actors. That type of acting would become quite common and even create its own niche in acting schools. Today, it is commonplace. Actors, sets, and locations would never be the same. Flying on dragons, commence.

Of course, a Spielberg film also includes masterful use of sound, among other things. *Jurassic Park* is no different. One of the most notable instances of expressive sound is the scene where everyone is stuck in their Jeep as the "ride" has broken down. The T-Rex pen looms, and unbeknownst to them, the electrified fence has been turned off (thanks to the diabolic plot by the evil scientist Nedry, played by Wayne Knight). (OK, he wasn't that diabolic and perhaps more of a bungling, greedy nerd. But he was kind of evil.) As the ride breaks down, however, everyone sits and waits for it start up again. Finally, in the distance, we hear the first footsteps of something big and powerful. Without the benefit of recording an actual T-Rex for the sound effect, the team settled upon the sound of a redwood tree crashing down. The resultant thud substituted quite nicely for the sound of the T-Rex. The *mise-en-scène* and sound then work together in the sequence to create one of the more memorable moments in the film. A harmless glass of water sitting in the cup holder sits motionless until the thud is hear in the distance. It ripples. More thuds and more ripples. The ripples become bigger and increase in frequency as the T-Rex gets closer and louder. The characters and the audience are similarly scared and anticipatory in a typical Spielberg scene that brings everyone to the edge of their seats. And while the sound of the footsteps was easy enough to find, the ripple effect turned to be one of the more vexing effects in the film. And we have Earth, Wind, and Fire to thank for it.

After trying every possible trick to get the ripples, Effects Supervisor Michael Lantieri got a call from Spielberg driving into the studio one day. Spielberg was yelling that after blasting Earth, Wind, and Fire through his car radio, the speakers were actually moving (OK, take a second to let that visual sink in). Lantieri then experimented with different sounds to get the effect, to no avail. Frustrated, he was relaxing at home that night and playing his guitar when he noticed his water glass ripple as he played. The next day, they tied a guitar string to the bottom of the water cup and ran it underneath the car. Someone plucked the taut string, and bingo! one of the more expressive effects in film history was born. On a film that featured groundbreaking special effects in both modeling and CGI, perhaps the most memorable effect was achieved by a guy lying underneath a car!

The music in the film is also classic Spielberg; famed composer John Williams wrote the score for the film, as he has for over 25 Spielberg films going back to *The Sugarland Express* and *Jaws*. It is one of the more fruitful and long-lasting partnerships in the history of cinema. Williams, who has been working in Hollywood since the '50s, is the classic of classic Hollywood cinema composers. Think the *Jaws* theme. Or Darth Vader's theme. Over his wonderful career, Williams has written some of the most memorable scores in history, and we hum his tunes long after leaving the theater and then some. In *Jurassic Park*, his score is typical Williams: the primary focus is "Theme from *Jurassic Park*" which we hear for the first time when we see the brachiosaurus. The Theme is imbued with feeling and emotion but also majesty and a soaring arc that matches the wonder of seeing dinosaurs for the first time. It is the high point of the film in the early stages and the music helps us get there. This is Williams at his best and Spielberg at his best and the two of them at their best together. That theme will come back throughout the film, in different iterations, and re-ignite those feelings when the film wants us to. It is classic Hollywood cinema at its best.

The sound includes more than score, and we also get ground sound effects—such as the simple footsteps of the T-Rex as he approaches the stranded cars. The sounds of the individual dinosaurs also stand out in the film, and the source sounds used for them also proved quite clever: the deep roar of the T-Rex (a Jack Russell playing); the bark of the Raptors (tortoises having sex); the hiss of the Raptors (particularly nasty geese); the gentle singing of the Brachiosaurus (donkey); the screech of the Dilophosaurus (snakes and hawks) as it kills Nedry, and the squealing of the Gallimimuses (horses in heat) as they stampede. Gary Rydstrom took a page from colleague and famed sound designer Ben Burtt and recorded natural and organic sounds for years before putting them together to form the sound world of *Jurassic Park*. Burtt became famous for the sound design of the *Star Wars* films, where he put together a whole new universe of sound and inspired a whole new generation of designers.

The cinematography and editing are more conventional classic Hollywood style: the cinematography gives us a plethora of establishing shots to allows us a sense of the environment, followed by a medium shot to set the scene, followed by close-ups to get us closer to the characters—both physically by being closer to the camera, but also psychologically closer to the subjects so that we identify and sympathize with them. This tried and true method of shooting scenes and sequences also allows for suspense and surprise as the audience is privy to information about which the character is not (the Raptors in the kitchen) and then complete surprise as when the Raptor stalks the game warden in a group. The cinematography allows for maximum tension and edge-of-your-seat-ness.

The editing aids in this projecting, simultaneously giving us long takes without cutting (anticipating the T-Rex arrival) that heightens our tension and keeps us on the edge of that seat in real time, along with faster cutting during the Raptor stalking scenes that moves the heartbeat needle while quickening the pace of the film. A Spielberg film is always about great pacing—he knows when to slow down and allow us to sit with the characters and then he knows when to move faster to give us the big action sequences we expect from him. Think *Jaws* and waiting on the boat. Or Indiana Jones in the cave. While neither the cinematography nor the editing are wildly expressive (as in *Children of Men* or *Upstream Color*, for example), the precision of both aspects, coupled with the classic Hollywood cinema aspects that both allows us to follow the action clearly without getting confused while also setting us up for the thrills, the film is extremely effective in its goals. All of these aspects are very much in the tradition of Steven Spielberg.

Perhaps the more interesting aspects of the film are in its themes, which are both timely and pertinent. The dominant subject of the Crichton source novel was genetic engineering, and the attendant theme was the dangers of unchecked genetic engineering and the hubris that leads us there. The 1990 novel was written in the context of genome mapping and research that started in the mid-'80s and was codified in 1990 by world governments. Cloning was also underway during this period, and Dolly the Sheep was finally cloned in 1996. The U.S. Supreme Court decision in 1980 allowing General Electric to patent genetically engineered bacteria also paved the way for GMOs and the proprietary nature of such engineering. Crichton put all of this together into an entertaining story that was timely and thoughtful, and Spielberg knew he wanted to adapt the book right away, acquiring the rights through Universal Studios before the book was even published. The science of the day was turned into science fiction.

That theme of genetic engineering gone awry shows up in the film as well, mostly in the first third of the movie, but it is a bit more diluted and eventually gets lost in the action of the final ⅔ of the film. One of the quirkier scenes from the film comes near the beginning, after the experts have arrived on the island and Dr. Hammond gives them a rundown of his research and his attraction. They discuss the ethical implications of engaging in such behavior, and it is Malcolm who voices the strongest ethical concerns about the engineering. But it is slightly glossed over in the film, whereas the book goes much further into the topic and the larger concerns. Hammond's presentation in the film version is also punctuated by an animated description and depiction of how, exactly, the genetic engineering of the dinosaurs is accomplished (more on that in the second part of this chapter). The animated sequence is then followed by the beginning of the attraction "ride" which moves directly into the lab. Of course, the scientists want to get a closer look, so they leave the confines of the ride and enter the lab. A baby Raptor is being born as they arrive, and it is quite remarkable. The film quickly moves on to the action from there. In the midst of this, the ethical aspects of the genetic engineering are lost, or at least forgotten. We come back to it at the end, when Hammond sees what destruction he has wrought. The moral may be tidied up a bit too much, however, and it all seems a bit too neat.

Unchecked capitalism is also a prominent subject, and the film perhaps spends more time here—the attendant theme being that not only does such a policy create winners and losers, it can also be downright dangerous and violent. The subject of capitalism and its attendant scary outcomes is a prominent focus of science fiction cinema and has been since its earliest iterations. The central focus of the first great science fiction film, *Metropolis*, was class division and the strife between upper and lower classes. In Fritz Lang's film, it manifests itself in the exploitation of the working class by the ruling class. The ideological sci-fi of the '70s was rife with such comments about capitalism, including, most notably, *Alien*. The mining ship Nostromo of the film is coming back from a deep space mining mission when it diverts to pick up the alien (at the behest of the corporation running the show). Aside from the race and gender divisions on the ship, the alien itself is a representation of pure capitalism: it is deadly, adaptable, it stamps out all competitors, and it is ruthless in the pursuit of its goals. The company who owns the ship finds the workers themselves "expendable," and the pursuit of the alien is greenlit because of its military and economic potential. We see similar topics in *Blade Runner* (1982) where economics is the driving force not only behind the division of who stays on the dying Earth and who leaves for newer off-world colonies, but also in the proprietary existence of the androids at the center of the film. The evils of capitalism have always been a stable and easy target for science fiction, and we see that in both the book and film versions of *Jurassic Park*.

Other prominent subjects come directly from Spielberg and his own imprint on the film and all of his films, for that matter. He has frequently worked with a book as source material, but he always makes the film his own and includes similar subjects and themes throughout his body of work. Prominent subjects in his films include the family as a disjointed or even fractured unit (in many cases, divorce), an absent father/father-figure, childhood trauma, and ordinary people forced into extraordinary action. We see many of these subjects, in one form or another, in his films, including *Jaws*, *E.T.*, *Close Encounters of the Third Kind*, and *The Color Purple*, among others. The fractured family in *Jurassic Park* comes through in Hammond's grandchildren—they are visiting because their mother and father are going through divorce. The kids are looking for a father figure, and Hammond is better suited for the role of grandpa and quite good at it, unless you consider the whole "let's put you in the path of a rampaging T-Rex followed by a scrum with deadly Raptors" thing. It thus falls to the reluctant Dr. Grant. The Dr. is not as reticent in the book as he is in the film, and Spielberg is able to personalize the characters and add just the right amount of *pathos* through Dr. Grant's relationship with the children. The final shots of the children with Dr. Grant is quite wonderful.

And therein lies the beauty/genius of Spielberg and the answer to the central question of Sci-fi: "What does it mean to be human?" *Jurassic Park* gives us a simple and quite lovely answer: to be human is *to be human* and not play God. The hubris of the scientists (along with the greed of the lawyer—who gets what is coming to him quite poetically in the shitter) is dangerous and not only functions as a cautionary tale but it also makes clear the danger it poses to the future, driven home pointedly by the travails of the children in the film. To be human is to take care of our children; to be human is to take care of environment. To be human is to be human.

—Vincent Piturro

Jurassic Park and the Dinosaur Renaissance

Joseph Sertich

The fast moving, birdlike dinosaurs of today's movies would have been impossible to imagine 50 years ago. For much of the 20th century, both scientists and the wider public largely reconstructed dinosaurs as dimwitted, slow-moving lizards. While many exceptions in popular culture and the scientific research attempted to change this narrative, for more than six decades, public perception was unwavering. Early flickers of today's dynamic dinosaurs appeared nearly as far back as their first recognition as a unique branch on the animal family tree in 1842. Early depictions of dinosaurs as ferocious, fast moving, and dynamic animals include a now iconic 1896 watercolor, *Leaping Laelaps*, by paleoartist Charles R. Knight, depicting two tyrannosaurs locked in combat. Teeth flashing and claws extended, the pair tumble across the Cretaceous landscape like a pair of big cats, one on its back and the other frozen in mid leap. But even this was an exception for Charles Knight, as he reinforced the slow-moving trope with giant sauropods depicted half submerged in swamps to support their immense weight, and most land-bound dinosaurs dragging their tails in lizard-like slow motion. By the 1950s, slow and stupid dinosaurs had been firmly entrenched in the public consciousness.

The first cracks in this narrative began with a discovery in the Montana badlands in 1964. Following earlier reports of productive dinosaur beds, Yale paleontologist John Ostrom uncovered a trove of fossils in the middle Cretaceous Cloverly Formation, among them a new raptor dinosaur he later named *Deinonychus*. Unlike the solitary remains of previously discovered meat-eating dinosaurs, the bones of several *Deinonychus* were found together, offering the first evidence for social behavior, maybe even cooperative hunting. *Deinonychus* was armed with large, birdlike claws on its feet, the largest of which was held off the ground by specially articulating toe bones. Its tail was reinforced by elongated projections from each vertebra into a stiff, counterbalancing rod that could only have been up held off the ground. Its muscular arms and clawed hands would have folded along the body in a posture very similar to the folded wing of a bird. Features like these, paired with Ostrom's research on the early bird *Archaeopteryx*, revealed striking similarities. Maybe dinosaurs were more bird-like than lizard-like.

Discoveries over the subsequent 20 years reinforced the active dinosaur hypothesis, often showing them to be much more dynamic and complex than previously believed. Dinosaur nesting and parental behavior was uncovered, again in Montana, in the mid–1970s by paleontologist Jack Horner. Nests, hatchlings, and adults of a new duckbilled dinosaur, *Miasaura*—the good mother lizard—showed that dinosaurs likely cared for and protected their nests; a much more bird-like behavior than expected. Ostrom's own graduate student, Robert Bakker, amassed a mountain of circumstantial evidence and creative drawings to argue that dinosaurs were not only active, they were mammal-like in their ecology and physiology, with a warm-blooded metabolism and corresponding activity levels. His popular book pushing these ideas, *Dinosaur Heresies*, burst onto the scene in 1986, inspiring a new view of dinosaurs among enthusiasts. Then in 1990, the science of dinosaur paleontology was splashed across the front pages of magazines and newspapers around the world with the discovery of a nearly complete *Tyrannosaurus rex* skeleton. The find, nicknamed Sue, entered a contentious legal fight that led to its confiscation in 1992 by the FBI. With the sudden, intense publicity, three decades of new dinosaur research demonstrating a novel, energetic view collided with the public's perception of dinosaurs as slow, gigantic lizards.

It was into this world, now known as the Dinosaur Renaissance, that *Jurassic Park* was released. Like the asteroid that knocked out the non-bird dinosaurs 66 million years ago, the film collided into the public consciousness, demolishing the last vestiges of the lizard-like dinosaur paradigm. In addition to envisioning dinosaurs as the dynamic, fast-moving bird relatives we now know them to be, *Jurassic Park* introduced and normalized many other emerging views of dinosaurs.

There is no doubt that the dinosaur stars of *Jurassic Park* were among the first to reflect the emerging views of the Dinosaur Renaissance. Across the board, the dinosaurs were shown as fast-moving, social, complex creatures on par with more familiar mammals and birds. At every opportunity, this point was hammered, from the steamy breath of a *Velociraptor* on a window, to the herds of duckbilled dinosaurs drinking from a lake, to the description of a group of running *Gallimimus* as a "flock." In *Jurassic Park*, the concepts of endothermic (warm-blooded), birdlike dinosaurs were presented to a global audience in entrancing, unforgettable detail. The depictions of four dinosaurs in particular, *Tyrannosaurus, Velociraptor, Dilophosaurus*, and *Brachiosaurus*, encapsulated two decades of dinosaur research better than any scientific paper or book, with far reaching repercussions still being felt today.

Among the first dinosaurs encountered in detail in the film is the immense *Brachiosaurus*, rearing up on its hind limbs to extend it long neck to reach a mouthful of leaves at the top of a tree. The feasibility of this maneuver, and the physiologic demands it would place on the skeleton and cardiovascular system, are still being debated today. However, it showed a dinosaur as a living, breathing creature going about its daily routine. The emergence of more *Brachiosaurus* from a lake in the distance, easily climbing onto the dry shore with tail held aloft, literally depicted the submerged *Brontosaurus* of Charles R. Knight leaving behind its watery prison. These dinosaurs, even at unimaginable sizes, were able to support their body on land as easily as any elephant or giraffe. Later depictions of *Brachiosaurus* as gentle "cows" browsing peacefully from treetops, and suffering from everyday ailments like a runny nose, powerfully unraveled the dinosaur-as-monster idea. Herbivorous dinosaurs were intelligent, peaceful giants, not bloodthirsty brutes.

The first encounter with the films superstar *Tyrannosaurus rex* (*T. rex*) introduced audiences to a scientific debate that was raging at the time: was *T. rex* an active predator or did its size limit it to scavenging other kills and carrion? Would the chained goat offered as bait be too easy and unappealing to a hunter triggered only by a chase? Throughout the film, this question is answered with force as a fleet-footed *T. rex* chases people, dinosaurs, and vehicles with the recklessness of a hungry lion. Depicted in a now widely accepted horizontal posture, tail held aloft to counterbalance the massive head, this was among the first exposure audiences had to a giant, bird-like dinosaur as envisioned by scientists. Only a few small physical details of the *T. rex*, like pronated hands, did not meet the scientific understanding of the time. Curiously, the cinematic detail that *T. rex* could only see movement did not reflect any established scientific insights of the time, and in hindsight now comes across as ridiculous. Given their close relationship with birds, carnivorous theropod dinosaurs likely had correspondingly sharp visual acuity. *T. rex* in particular, with forward facing eyes similar to birds of prey, had well-developed depth perception, able to judge the distance and movements of prey with precision. Healed bites on the bones of likely *T. rex* prey like *Triceratops* reveal that *T. rex* attacked and hunted its prey. Though unlikely to turn its nose up at an easy meal of carrion, the movie version of *T. rex* as an active, bird-like hunter hits the nail on the head. The film's depiction of *T. rex* as a quick ambush predator, particularly when she leaps from the forest edge to surprise a stampeding "flock" of the ostrich-like *Gallimimus*, is precisely the version of *T. rex* that has persisted to this day.

Among the most unusual of the movie's dinosaurs is the carnivorous *Dilophosaurus*. Numerous liberties were taken in the depiction of this particular dinosaur, more than any other, but it did introduce audiences to new, emerging concepts in dinosaur physiology and appearance. *Dilophosuarus*, known from Early Jurassic rocks in the American West, would have stood up to eight feet tall and measured more than 20 feet long, much larger than its diminutive cinematic version. It is the addition of an extendable neck frill and sticky venom that veers most radically from scientific reality, but also reflects new thinking in dinosaur appearance and behavior. With the exception of very rare soft tissue preservation, nearly everything we currently know about dinosaur anatomy and behavior is based on skeletal anatomy and trace fossils. Evidence for a fleshy neck frill, or venomous spit, have never been found in *Dilophosaurus*, or any other dinosaur for that matter. However, these features did open audiences to new possibilities, expanding reconstructions of dinosaurs beyond the dull conservative norm. A new world of speculative dinosaur anatomy has blossomed since the early 1990s, though much of the speculation is grounded in comparisons with living animals. The bright colors of the Dilophosaurus frill reflect modern insights into probable dinosaur color vision and use of color for communication, display, and warning. Scientists have also begun carefully searching new and old finds for clues about dinosaur appearance and behavior with spectacular results. In some extraordinary cases that may not be as rare as first thought, soft tissues are preserved. We just needed to slow down and get creative to find them.

Undoubtedly the star dinosaur of the film, *Velociraptor*, also had its share of creative liberties taken, with the most obvious being its size. The dinosaur *Velociraptor*, a near relative of the dinosaur that kicked off the Dinosaur Revolution, *Deinonychus*, is known from the desert sandstones of the Gobi Desert in Mongolia. Unlike its more famous movie namesake, actual *Velociraptor* adults were much smaller, likely standing no more than three feet tall, about the size of a turkey. The huge, five-and-a-half-foot tall *Velociraptor* in the film was an intentional exaggeration that exceeded the maximum known size of any raptor dinosaur at the time. That is, until a fortuitous discovery during production saved the day. The remains of another close relative, the six foot tall *Utahraptor*, were discovered during production and ushered the movie's giant raptors from science fiction to science fact just in time for the film's release. Many of the other characteristics of the movie's raptors were based on recent scientific discoveries and may have been the first time they were encountered by wider audiences. Bird-like behaviors, stiff tails, and cooperative "pack" hunting depicted in the movie were derived from the groundbreaking work of John Ostrom three decades earlier. However, a few key details were left out, including the bird-like hand position. In order to get raptor hands to work with door handles, the wing-like folding described by Ostrom had to be modified to a more human-like pronation. Regardless, the *Velociraptor* of *Jurassic Park* introduced audiences to a truly terrifying new picture of intelligent, fast, cooperative dinosaurs. The old, slow lizards of the early 20th century were officially extinct.

In addition to its revolutionary recreation of dinosaurs, the film's depiction of cloning was well ahead of its time. Unfortunately, many aspects still exist too deeply in the realm of science fiction to expect the return of dinosaurs. Just three years after the film's release the world was introduced to the first cloned vertebrate, Dolly the sheep, and it seemed cloned dinosaurs were just around the corner. Indeed, in the decades since, cloning has become routine and vastly improved through advances in molecular engineering. The recent CRISPR revolution, allowing unprecedented genome editing, has the potential to bring the type of gene editing and cloning depicted in the film to reality. However, the main premise of the film, the discovery of intact dinosaur DNA trapped in fossil amber for millions of years, is today considered virtually impossible. Amber, preserved tree sap, is relatively common in dinosaur-bearing rock units around the world. In fact, over the past several years, unprecedented discoveries of fossil amber from the Cretaceous Period, dating at more than 90 million years old, have been made in Burma with the type of miraculous preservation one might hope for. Rare examples preserve fossil animal remains including dinosaur feathers and even a section of a small dinosaur tail. Despite their exquisite preservation, freezing details down to the cellular level, amber is a terrible medium for preserving fragile DNA molecules. In fact, DNA is so susceptible to breakdown due to temperature fluctuation and oxidization that it is highly unlikely that anything other than the smallest of DNA fragments from a dinosaur genome have the potential for long term preservation over millions of years. Indeed, the Burmese amber specimens preserve beautiful impressions of extinct animals and remains of bones and proteins, but no genetic molecules. Even if we did miraculously recover enough genetic material to attempt dinosaur cloning, scientists would face nearly insurmountable challenges creating a viable embryo, let alone bringing it to the type of successful hatching depicted in the film. While bringing dinosaurs back from extinction may not be possible, the potential for bringing back recently extinct organisms like the Tasmanian thylacine or the Passenger Pigeon, or even more distantly extinct Ice Age animals like the wooly mammoth, are becoming more and more realistic. While we may never have a Jurassic Park, de-extinction could bring us a Pleistocene Park complete with Ice Age wolves, lions, mammoths, and rhinos reclaiming the tundra of Siberia.

The Dinosaur Renaissance as first revealed to global audiences by *Jurassic Park* has only accelerated in the decades since the film. Inspired by new advances in the field of paleontology, renewed interest in the science, and a new unencumbered generation of paleontologists, dinosaur paleontology continues to surge ahead. Despite being on the cutting edge of dinosaurs nearly three decades ago, our picture of dinosaurs has changed dramatically as new fossils have come to light. Three years after the release of *Jurassic Park*, the world was introduced to the first feathered dinosaur. Small theropod dinosaurs, many times closer to birds in the family tree than to their horned and armor-spiked cousins, were discovered in fine lake sediments deposited at the bottom of an ancient lake in China nearly 100 million years ago. The first of these discoveries, a small carnivorous dinosaur named *Sinosauropteryx* in 1996, revealed that the similarities between birds and dinosaurs extended beyond just boney morphology. In the decades since, a trove of new fossils from China and elsewhere have piled on the evidence for feathered dinosaurs. Even *Velociraptor*, though never found preserved with feather impressions, shows the boney landmarks on its forearm of anchor points for feathers. The evidence for quill-like structures has even shown up on the herbivore side of the dinosaur tree, with small bipedal ornithopod dinosaurs and horned dinosaur relatives preserved with stiff, porcupine-like quills, possibly a type of proto-feather. Cutting-edge investigations of these feathers under high-powered magnification has even begun to reveal dinosaur colors. Based on the work of Ostrom, and combined with new discoveries and analyses, we now know that living birds, in all of their resplendent plumage, are a surviving dinosaur lineage. Therefore, it should not come as a surprise to learn that dinosaurs were decorated with patterned and colored feathers of all shapes and sizes. So pervasive is the feathered dinosaur phenomenon, the *Velociraptor*, *T. rex*, and even the *Dilophosaurus* of *Jurassic Park* now look naked in hindsight. A remake of *Jurassic Park* would certainly need to embrace the inner bird of dinosaurs and cloak them in colorful plumage.

Among the greatest influences of *Jurassic Park* on the scientific community is the creation of the *Jurassic Park* generation. Children and young adults exposed to *Jurassic Park* fully embraced the ideas of the Dinosaur Renaissance as a new dogma. This included a new wave of young paleontologists unencumbered with stale views of dinosaurs as slow, dimwitted lizards, and armed with new technology. New discoveries of dinosaurs around the world have accelerated in recent decades as children inspired by *Jurassic Park* pursued careers in paleontology and pushed the boundaries of the science. New species of dinosaurs are being discovered and named at a record pace from every corner of the globe. In addition, new insights into their behavior, ecology, appearance, and physiology have accelerated. We now know that dinosaurs were a part of dynamic ecosystems with similarities to and differences from today's world. As we push wild places, and wildlife, to the precipice of a new mass extinction, we can look to the past, to dinosaurs, to understand how the world around us will change. Perhaps dinosaurs can help us avoid their same fate.

Monstrous Life Finds a Way: *Jurassic Park* and Monstrosity

CHARLES HOGE

They're not monsters.

Or are they?

During a lull in the action in *Jurassic Park* (1993), and shortly after a narrow escape from a rampaging *tyrannosaurus rex*, siblings Lex (Ariana Richards) and Tim (Joseph Mazzello), along with the reluctant protector Dr. Alan Grant (Sam Neill), take refuge in a tall tree where they are able to watch a gigantic brachiosaurus grazing among the high branches. Lex is initially terrified, but her brother attempts to calm her by informing her that this dinosaur is not carnivorous and that, of the park's dinosaurs in general, "they're not monsters. They're just animals." Grant adds that "they're just doing what they do."

But, in "just doing what they do," are the resurrected dinosaurs that escape from their containment and stalk the grounds of Jurassic Park for human prey more than animals? Are they actually monsters?

It's a fair question.

And in the most direct sense, the answer is yes. The carnivorous dinosaurs in particular are unquestionably monstrous, morphologically speaking; the razor-sharp dentition of the tyrannosaurus rex, the poison-spraying fanged jaws of the *dilophosaurus*, and the viciously-curved talons of the velociraptors leave no doubt as to their dietary intentions and the potential to be dangerous for human beings. Furthermore, these dinosaurs' tendency to attack humans, just as they might attack any other prey species, makes them monstrous in their behavior as well. In fact, when we recall how the t-rex relentlessly chases a speeding jeep and how the velociraptors strategically stalk the children through the park's giant kitchen, we become aware that these dinosaurs demonstrate a tenacity uncommon in natural predators in their continuous, determined pursuit of human prey. With apologies to Tim, it must be pointed out that this is *not* actually what animals do, normally. (Maybe more important, though, is the recognition that the dinosaurs do not respect the self-proclaimed status of the human as the king of nature, as these predatory dinosaurs do not perceive us as special, alone at the top of the food chain that we created: to the t-rex in particular, we are really just another item on the menu.) David Gilmore, in *Monsters: Evil Beings, Mythical Beasts, and All Manner of Imaginary Terrors*, undertakes the difficult procedure of defining monsters, and his eventual definition acts as a virtual checklist as it seems to describe the carnosaurs of Jurassic Park perfectly: he offers that monsters are "terrifyingly large, awe-inspiring in their destructive powers, retrograde or devolved in conception, toothy, gratuitously malevolent toward humanity, and of course hungry for human flesh."

But a monster is also monstrous because of *where* we find it. Any sort of entity that is out-of-place is essentially a monster because it wreaks havoc on our expectations by showing up where it shouldn't; in other words, it is anything that we encounter in places in which it isn't expected to be found. A dinosaur finding a way to survive in our modern world, a good 65 million years after the fossil record tells us they all went extinct, provides a productive example of this kind of monster. And among cryptids (creatures undocumented by empirical science but witnessed in the wilderness by people nonetheless), the shape of the dinosaur is remarkably popular; whether it is the *plesiosaur*-like Loch Ness Monster of Scotland, the giant, long-necked saurian-type Mokele-Mbembe that inhabits the watery regions of the Congo River Basin, the occasional *pterodactylic* creature flying over the midwestern United States, or even the 3–4-foot-tall mini-*t-rexes* that might haunt Canyonlands National Monument in Utah, these apparent "survivals" are monstrous because they quite simply should not exist. Maybe more significantly, their impossible presence undercuts our confidence in our own knowledge of the world around us: if we believe that dinosaurs are extinct, yet we continue to see them in our world, then we are clearly not as informed as we think we are. One of the linguistic roots of the term "monster" is *monstrum*, which may be translated from Latin as "that which warns." The dinosaur that refuses to be extinct is warning us, in its monstrous fashion, that our understanding of the natural world, and our placement within it, is fundamentally flawed. Even further, the presence of dinosaurs challenges humanity's self-appointed status as the ruling species on earth, as *Jurassic Park* demonstrates (building on a wealth of earlier dinosaur films as it does so) just how outmatched we are, and how truly unworthy an opponent we are when we are facing the monstrous "thunder lizards" of the deep past.

But the dinosaurs of *Jurassic Park* are not unexplainable animals in the manner of the Loch Ness Monster and its cryptid kin. Their origin can be found within the film's (and of course the book's) storyworld. They have been genetically engineered, cobbled together by prehistoric DNA which has been extracted from ancient mosquitoes trapped in amber, "repaired" by fusing it with DNA from extant animals, and finally gestated and hatched in carefully-maintained laboratory settings. So the very safety that tens of millions of years has granted the human species by separating us from these forbidding monsters is defeated by human endeavor, as Jurassic Park scientists reach across uncountable prehistoric fathoms of time to bring the long-gone monsters directly to us.

The Dinosaur as a Monstrous Construction

According to Jeffrey Jerome Cohen's influential "Monster Culture (Seven Theses)," one of the most significant factors that identifies a monster is that it is culturally constructed. In other words, the monster is "given life" by the human imagination and more specifically, what we are afraid of or nervous about on a cultural level. So it is we, as participants in human culture, who make monsters. Without human beings to channel their cultural fears into monstrous shapes, creating frightening symbolic repositories for their own fears and anxieties, there would be no monsters. (In this way, it may be useful to consider Frankenstein's monster in Mary Shelley's 1818 novel *Frankenstein*: the monster is often interpreted by literary critics as a symbolic avatar for Shelley's culture's fears that scientific advancements were allowing human beings to "play God" and that such transgressive endeavors might disrupt the coherent line that separates life from death, which would bring chaos, horror, and ruin to the world as a result.) The dinosaurs of *Jurassic Park* are literal constructions, scientifically reengineered by InGen, a corporation that uses science to monetize resuscitated dinosaurs for a culture that is fascinated with, and will pay heartily to see, them. These lab-resuscitated dinosaurs are positively Frankensteinian, and their monstrous behavior reflects the cultural anxieties that the late 20th century shared with Shelley's time: monsters are the inevitable result when overly ambitious science fails to consider the consequences of its actions. So these dinosaurs are the very monstrous product of the kind of thinking that pushes human knowledge recklessly into dangerous, ultimately disastrous territory.

Chaos awaits such endeavors.

The specific cultural fears that these resuscitated prehistoric monsters intend to reflect is essentially explained by Jeff Goldblum's "rock star" Chaos Theorist Ian Malcolm. The underlying horror, he explains, is that our scientific knowledge may advance to the point that we no longer ask whether we *should* do what we become capable of doing, and that we may be punished for our transgressions by the very things we transgress to create. (This is hardly a new fear, as we can recall *Frankenstein* again, and how it articulates the same concerns in its presentation of the ambitious Victor Frankenstein, who creates life from death without hesitation, then is haunted and ultimately destroyed by the very creature he creates.) Malcolm riffs "God creates dinosaurs. God destroys dinosaurs. God creates man. Man destroys God. Man creates dinosaurs," and Ellie Sattler (Laura Dern) punctuates this by reminding him that "dinosaur eats man" and, what's more, "woman inherits the earth" (a poignant statement when one remembers that the entire process of engineering dinosaurs at Jurassic Park involves the erasure of feminine power: creating and controlling female dinosaurs and eliminating biological motherhood as a possibility).

Malcolm warns John Hammond, the enthusiastic head of InGen, that "the kind of control you're attempting is not possible. If there's one thing the history of evolution has taught us, it's that life will not be contained. Life breaks free. It expands to new territories. It crashes through barriers. Painfully, maybe even … dangerously, but and … well, there it is." In other words, a person may create something or cause it to come into being (this is a vast generative spectrum, and it includes giving birth to a child, engineering a dinosaur, painting a mural, writing this article, etc.) but can have no control over that creation as soon as it is released (or escapes) out into the world. This is how Chaos Theory works, according to Malcolm, and, interestingly, it shares a lot of conceptual territory with general theories of how monstrosity works as well.

The monster, according to Cohen, will always "break free" and "crash through borders." Fundamentally, monsters do not respect boundaries of any kind, and will always escape any attempts to contain them. Jurassic Park's dinosaurs escape containment of every kind. Physical barriers, which are constructed and maintained in order to protect human tourists and workers from the more dangerous dinosaurs are all seen to get knocked offline during the course of the film, for example. Scientifically-constructed boundaries are erected, which attempt to control the sex-expression of the dinosaurs, engineering their genetic codes to produce only females so that they can only reproduce in the lab by way of human action; however, "life finds a way" and at least some of the dinosaurs mutate into males in order to allow for natural reproduction to become possible. These dinosaurs even defeat rhetorical barriers. It should be added that even the name "Jurassic Park" carries hints of a thing that cannot be contained, as several of the most prominent dinosaurs in the park (namely the *t-rex* and the *triceratops*) are not from the Jurassic Era at all (which sprawled from about 200 to 145 million years ago) and are rather denizens of the later Cretaceous Era (which spanned from the end of the Jurassic to 65 million years ago). On a subtle level, this reminds the viewers that these terms have no real meaning beyond that which our culture has ascribed to them. In every discernible way, we can see how life, or in this case monstrous life, "cannot be contained."

It should be noted that the ability of the dinosaurs to escape containment comes as a result of human action. The electrified fences that contain the t-rex are shut down by Dennis Nedry (Wayne Knight), a shifty InGen employee who needs to disable the security system so he can steal some dinosaur embryos to sell to his employer's competitor: human greed, coupled with the financial value that humans have assigned to the dinosaurs they can reengineer motivate the factors that allow for the *t-rex* to escape containment, not any innate monstrosity within the *t-rex* itself. Later, as technician Ray Arnold (Samuel L. Jackson) resets the security system, a brief power lapse permits the velociraptors to escape as well. But it's not just the physical barriers the dinosaurs can overcome, as they also break free of the boundaries imposed by the human-designed engineering process. Specifically, the frog DNA which was used to fill in the gaps in the dinosaur's recovered genetic material opened the gate, so to speak: when it expressed itself genetically, the frog genes allowed some of the all-female dinosaur population to mutate into males, thus defeating the human attempt to control their reproduction.

None of these boundary-violations came about from any directly monstrous dinosaur activity. They were all indirectly the result of human efforts. One may wonder, though, if these actions might have subconscious underpinnings and that, just maybe, these humans wish the dinosaurs to break free. Such actions, if motivated without conscious awareness, suggest the consideration of another element of Cohen's monstrous definition, which involves the tendency of monsters to represent, even as they terrify us, a kind of desire or longing. We want to be like them: free, powerful, unfettered by societal restrictions and rules of decorum. Many of us think that it would be fantastic to be a dinosaur, in other words, and releasing them from Jurassic Park allows us, vicariously, to attain that same kind of monstrous freedom. But there may be even more at work here than a subconscious desire for us to experience what it's like to be monsters-by-proxy. Elizabeth Kingsley, in her blog "And You Call Yourself a Scientist!," expresses the deep-seated cultural love we feel for dinosaurs and is worth quoting in full: "The relationship between modern man and these extraordinary, long-extinct creatures is a peculiarly emotional one. There is a strange, deep yearning associated with it, one that compels small children to wrestle with multisyllabic names; that sends adults out into the wilds of our world to sift painstakingly through its dust; that turns theoretical musings into violent blood-feuds; and that, yes, inspires writers and film-makers to create, and makes the general public eager to consume."

And perhaps even the Park itself is designed to reflect a part of this deeply-entrenched love for the monstrous versions of dinosaurs that stalk our imaginations.

Prehistoric Thunderdome

In our cultural imagination, the prehistoric world is a terrifying arena full of ever-battling monsters, perhaps best typified by the ubiquitous presence in children's dinosaur books of the tyrannosaurus rex fighting one in a series of heavily-armored dinosaurs (most often the *triceratops* or a similar species, but occasionally an *ankylosaurus* or even a *stegosaurus*). The duel to the death between the slavering carnosaur and the tank-like herbisaur was a staple scene in early cinematic depictions of dinosaurs; one thinks maybe most strongly of Disney's animated epic *Fantasia* (1940) and its depiction of a horrifying, red-eyed allosaurus (frequently considered to be a *tyrannosaurus*, though its elongated, three-fingered arms disqualify it from such an identification) fighting and eventually killing a valiant *stegosaurus*, all while Stravinsky's "The Rite of Spring" played furiously in the background.

Stop-motion animation pioneer Ray Harryhausen produced several of the most iconic reenactments of this trope with his choreography of intense fight scenes between a putative allosaurus, the Gwangi, and a *styracosaurus* (an elaborately frilled relative of the triceratops) in *The Valley of the Gwangi* (1969) and a *ceratosaurus* and *triceratops* in *One Million Years B.C.* (1966). (This battle is especially notable because the *triceratops* wins!) Proving that predatory dinosaurs would take on all impressive opponents, even if they weren't dinosaurs, we recall the *t-rex* that appears on Skull Island in *King Kong* (1933) to fight the giant ape. And the trope goes back further, to the 1925 *Lost World*, in which an animatronic *allosaurus* and *t-rex* stop-motion rampage through the film, attacking and fighting every other dinosaur they encounter. At the dawn of the cinema age we can find the first (existing) battle between a stop-motion *t-rex* and a *triceratops*, in Willis O'Brien's 1918 *The Ghost of Slumber Mountain*.

But this notion of a prehistoric world as a place of constant dinosaur-on-dinosaur violence predates the cinema and can be found lurking within the halls of our most esteemed museums, too: Charles Knight's 1927 iconic panoramic mural, hanging at the Field Museum of Natural History, depicting a *tyrannosaurus rex* and *triceratops* facing off and preparing to do battle, cemented the notion that these two impressive dinosaurs necessarily squared off as mortal enemies. And the trope may be discovered even deeper in our cultural past, stalking the Victorian imagination, too. Louis Figuier's print "The *Iguanodon* and the *Megalosaur* (Lower Cretaceous Period)" (1863), for example, depicts two viciously-fanged quadrupedal dinosaurs looking evilly pleased as they chew on one another. And John Martin's 1840 mezzotint "The Sea-Dragons, as They Lived" (for the frontispiece of Thomas Hawkins' *Book of Great Sea-Dragons*) reveals a moonlit scene in which at least three different kinds of demonic-looking prehistoric marine reptiles, one mosasaur and a pair of *plesiosaurs*, churning in a turbulent black sea, are engaged in a wildly ferocious battle, directly next to a collection of razor-toothed *pterodactyl*-like creatures greedily devouring the carcass of what looks a bit like an *ichthyosaur*.

The point here is that we have, for quite some time, envisioned the prehistoric world as being monstrous: it is a violent landscape in which hideous monsters constantly do bloody battle with one another. This image of the dinosaurs' world as one red in tooth and claw served Victorian fantasies of superiority, as well: "just look at how far we have come," a glimpse into a frightening (and totally constructed) ancient world allows us to say to ourselves.

So this monstrous world, in which titanic, immeasurably powerful monsters engage in ceaseless fights to the death, is pretty much the world of Jurassic Park as soon as the system goes offline and the *carnosaurs* run free. The battle between the *t-rex* and the *velociraptors* at the film's conclusion demonstrates clearly how much this trope continues to live and breathe in Jurassic Park. Chaos Theory, according to Malcolm, tells us that this is inevitable, and monster studies enthusiastically supports this claim: monsters absolutely cannot be controlled. But, in some way, isn't that the appeal of Jurassic Park? Would human culture be as fascinated with dinosaurs if they could truly be contained?

The Monstrous Imagination

Jurassic Park is the human imagination made flesh, and its representation of a "lost world" borrows very strongly from the Victorian fantasy depictions of constant, spectacular dinosaur warfare. But until Chaos has its way, it is a dormant horrorscape of the imagination, waiting to be activated. Fortunately, Chaos has been encouraged at every step, as the island is populated by dinosaurs that have been genetically inscribed with the potential to become monsters, and which are spurred further along in the direction of monstrosity by all sorts of human folly. In the same sense that nature cannot recreate the extinct dinosaur on its own and requires human assistance in that direction, the revitalized dinosaur cannot become a monster without the human imagination to make it so. And the human imagination is all too willing to cooperate.

Ultimately, the dinosaurs of *Jurassic Park* do seem to be monsters, because that's exactly what we need them to be.

Chapter 9: *King Kong*

Kong atop the Empire State Building in *King Kong* (RKO Radio Pictures, 1933).

  Merian C. Cooper was an interesting and flamboyant character, even in the wild, early days of Hollywood. He was born in Jacksonville, Florida, in 1893, his birth coinciding with the birth of cinema (1895) and the birth of flight (1903) soon after. He grew up in a world of wonder, adventure, and burgeoning technology, and he was thrilled by it all. He chased Pancho Villa as a member of the Georgia National Guard, he was a pilot in World War I (where he was shot down), he fought in the Polish Air Force, he was a Russian prisoner, he became a reporter, he traveled the world, he helped to form Pan Am, and then he made his way to Hollywood, producing films for RKO with famed producer David O. Selznick. He began his filmmaking career producing and directing documentaries he shot while traveling the world. His second feature film as producer and director, however, would be one of the most thrilling, innovative, and technically difficult films in the history of cinema. He said that he hatched the idea after having a dream of a giant ape terrorizing New York, and then imagined the ape at the top of the newly constructed marvel, The Empire State Building (finished in 1931). From there, he built a wild and imaginative story that he simple called "Kong." David O. Selznick changed that to *King Kong*, and it would go to become one of the most famous and iconic films in history.

From the beginning, Cooper imagined his film as a cross between some of the exotic documentaries he had shot, and a monster movie. Set in Depression-era New York City, the film takes its time getting to the action: as adventure film director Carl Denham (Robert Armstrong) is about to set sail on a mysterious film shoot, he is without a leading lady. The night before he leaves, he finds young Ann Darrow (Fay Wray) alone and hungry on the street, and he talks her into going on the journey. While on the ship during the journey, the first mate John Driscoll (Bruce Cabot) falls in love with her. Nobody knows where they are going, but as they get closer, Denham tells everyone they are headed for the mythical Skull Island—a remote place only passed down in legend and rumored to be the home of a giant ape known as "Kong." The crew finds the island, sets ashore, and discovers a tribal ceremony with the natives preparing a young woman as a sacrifice to Kong. But after seeing the "golden woman" Ann, the natives steal her from the ship that night and offer her to Kong. Denham and crew pursue, fighting off dinosaurs (!) and Kong himself. Finally, Driscoll saves her (after Kong has fallen in love with her as well), and they retreat to the ship. Kong follows, breaking down the giant wall built to keep him out and rampaging through the village eating and stamping villagers along the way. The crew is able to subdue him with "smoke bombs" and bring him back to New York, where he is chained up on a stage for a live audience. After seeing Ann with John and irritated by the flashes of the photographers, Kong goes wild and breaks free, terrorizes New York City, steals Ann, and climbs to the top of the Empire State Building, where he is shot down by planes (one of the pilots being Cooper himself). Ann is saved.

The plot is a lot to take, especially for the pre–Hays Code Hollywood, just a few years after talkies were invented (the first sound film was in 1927) and considering the rudimentary special effects of the era. There are several historical, socioeconomic, and filmic contexts important for better understanding the film and affording it its proper due and situating it in terms of those contexts. The first context concerns the era in which the film was made—the apex of the great Depression in 1933. The country had been devastated for several years, and depression (small "d") had really set in. The opening of the film speaks to this dynamic, as Denham laments over not having a leading lady the night before he sets sail out of New York Harbor. He relates this fact to the Captain, and then makes one last ditch effort on the streets of New York before leaving. He immediately encounters a line of women waiting in the cold for food and lodging that night. Soon after, he finds Ann as she is caught trying to steal food from a bodega. She is without family or anyone else to help her, and Denham buys her a meal and pitches his adventure to her. The movies, essentially, "save her." There is quite a bit to unpack here in just this opening, with shades of Charlie Chaplin in the 1910s and even the earlier days of filmmaking as well.

Charlie Chaplin made a short film called "The Immigrant" in 1917, just as he was becoming the biggest star in the world. The story was semi-autobiographical, about a European immigrant coming to America in the early 1900s. Along the way, he befriends a woman and her elderly mother on the ship, and he also gambles his way to making a little money (representing the danger/gamble of coming to the new world penniless). He and the other immigrants are also herded—like cattle—behind a rope as they sail into Ellis Island past the Statue of Liberty. A cut from Charlie admiring the Statue of Liberty back to him stung by being roped-in makes clear the point: the arrival in the Land of Liberty doesn't match up to the myth or the great symbol of the Statue. He kicks an immigration officer in the butt as the man walks away—likely drawing laughs from the mostly non-English speaking immigrant crowd of early cinema. But that kick would later get him into trouble with J. Edgar Hoover, who accused Chaplin of being a Communist (and therefore anti-American) and denying him re-entry into the country in the '50s. Later in the film, Chaplin and his friend from the ship meet an artist who hires them to be models. Chaplin is "saved" by art, just as he was in real life. In *King Kong*, art not only saves Ann as it did Charlie, but it also gave their audiences an escape and some hope that their lives in America would get better—just as it did in the teens, before the prosperity of the '20s. But Ann's path becomes a little different than Charlie's, which leads us to an unsavory reading of the film.

Ann is a single woman in the Depression, a terrible fate in a world where women didn't work and if they did, they were abused and exploited. The line of women at the beginning of the film provides the evidence (almost documentary-like, with shades of Cooper's own documentary experience). While everyone was affected by the Great Depression, women, and in particular, single women, were much worse off. Denham plucks her from this awful fate and brings her along on the ship, where she also falls in love with first-mate Driscoll. This is where the racism of the film starts to trump the sexism: she is captured by the (black) natives and offered up as a sacrifice to the giant (black) ape. The ape, brought back from a far-away, exotic island, then terrorizes New York City until he is bested by the white males. Ann is also saved by Driscoll, and we assume they live happily ever after. The allegory here is clear: the dark creature (an ape, conjuring racist images of black people in a very segregated society only some 60 years removed from the Civil War) comes to New York to take the (white) women and brutalize the city. The white men save day, and as for the women, well, of course the only way you will be safe in the world of the '30s is to find yourself a man. The saddest part of the story is that both may be true: the country was extremely racist and would have been soothed by the narrative undertones, and, a woman would have been safer being married during that time. Strangely, however, there is also another reading to the film that shows the Imperialist and unseemly nature of the Americans who do their own brutalizing of the people they shackled and chained as they forced them into our country (for the sake of pure profit). Perhaps the biggest victory of the movie is to humanize Kong and makes us sympathetic to him, even lamenting his death at the end. Even still, on another level (discussed in a moment), the ape is either a rubber suit, an animated puppet, or a screen projection of one or the other. The film makes us *feel for this being*, showing the similar exploitation of millions and how our own greed devastated them. It's such an interesting film that either reading (or both) may be true. The multitude of interpretations have been discussed for decades, and continue to resonate today.

Another context of the film concerns the film business itself and a changing morality therein. The roaring '20s were just that in the movie business—a wild time where Hollywood became "Hollywood" and newly-rich stars became the kings and queens of the burgeoning business and the rapidly growing metropolis of Los Angeles. Stories of debauchery and hedonism were the stuff of legends, with a dark side to that as well: in 1921, comic megastar Roscoe "Fatty" Arbuckle was accused, and later acquitted, of manslaughter (ruining his career and reputation); director William Desmond Taylor was found dead and rumors abounded about his killer; and in 1924 producer Thomas Ince was killed on a yacht owned by media magnate William Randolph Hearst (think *Citizen Kane*) with Charlie Chaplin and Hearst's mistress Marion Davies (with whom Chaplin was rumored to be having an affair) aboard. Ince's murder was never solved (like Taylor's), and money/fame/press were most certainly a factor in it all. The sum total of these and other salacious events was of a Hollywood that was essentially lawless and that the increasing sex, violence, and hedonism in the movies themselves was getting out of control (especially during a period in the late '20s and early '30s dubbed the "pre–Code talkies"). Facing possible censorship from the government and the backlash of an increasingly prurient public, Hollywood decided to censor itself. The big studios decided on William Hays, a former Postmaster General and well-respected citizen (read: pious) to formulate and institute the Code—formally "The Motion Picture Production Code" but colloquially known as the "Hays Code." The Code would be in effect from 1934 until 1968, and it would not only change the course of the industry in the nascent sound era, it would inform filmmaking for decades. The Code would restrict filmmakers in many ways, and especially coming when it did—at the beginning of the sound era—it would set the industry and art back for many years. Aspects of the Code include the following: "No picture shall be produced that will lower the moral standards of those who see it. Hence the sympathy of the audience should never be thrown to the side of crime, wrongdoing, evil or sin; Correct standards of life, subject only to the requirements of drama and entertainment, shall be presented; Law, natural or human, shall not be ridiculed, nor shall sympathy be created for its violation." The effects were far-reaching: there was no explicit sex, violence, or drug use; women and men could not be shown in the same bed (even married couples; think "I Love Lucy" beds); kisses could last for no longer than three seconds; and toilet bowls were certainly not allowed. Film had become chaste.

While the Code did not go into effect until 1934, it loomed over the production and release of *King Kong* in 1933, and it would illuminate the move from the looser restrictions of pre–Code Hollywood to the highly restrictive post–Code era. The original release version included scenes that were only viewed for a short time and would be censored upon later releases: in one scene, Kong undresses Ann as he holds her, smelling his fingers afterward; another scene has a Brontosaurus viciously mauling several of the crew in a lake; later, Kong chewing on villagers on the island as well as several New Yorkers later in the film; Kong smashing villagers with his feet and mushing them into the ground while on the island; Kong blithely dropping a woman to her death on the side of a NY building; two shots of Ann wearing a see-through shirt without a bra; and a scene with animals and insects swarming the ship's crew during another sequence on the island. Only those who saw the film upon its initial release would view these scenes; when the film was re-released in 1938 (and several more re-releases right up to the '70s, including a TV run in the '50s and '60s), it would be heavily censored. The uncensored version shows the brutality and violence of Cooper's vision, and it makes the film even more frightening while highlighting the transition to the Hays Code-era. The Code would be replaced by the ratings system in 1968, opening the flood gates for sex, violence, and whatever else Hollywood could dream up. *King Kong* is thus a fascinating case study in the transition to the Code-era films and beyond.

Yet even considering all of the controversy surrounding the film throughout the decades, the film has endured for many reasons, perhaps the biggest of which is how the film changed the face of filmmaking forever, for the same reasons *Jurassic Park* did in the 1990s: special effects. *King Kong* set the standard for several different aspects of special effects in film, many of which had been used before (in bits and pieces) and would become perfected in the film, and other techniques that were invented for the film. *King Kong* put everything together into one marvelous package and would become a textbook and source material for special effects professionals for decades to come.

The history of special effects in cinema goes back to the earliest days of the art. One of the first great filmmakers, Georges Méliès, used the stage as the basis for his films; they were basically filmed stage plays and based on various sources. He came to the cinema as a magician in the late 1800s and starting making films soon after the Lumière Brothers premiered the first films in 1895. He infused his love of magic into film and would be the first special effects master (his most famous film is "A Trip to the Moon" from 1903, showing, you guessed, a moon shot). For an underwater shoot, for example, he would place the camera in front of an aquarium and then add the action on the other side of the aquarium, or, do multiple exposures to add effects: he would simply rewind the camera and add something to the original shot. Or he would make things disappear by stopping the camera, changing something in the placement of the shot, and then re-starting the camera. Later filmmakers would build on these techniques and go further still. Painted backgrounds were added on sets; animation began in 1908 in France and quickly moved to the United States; and filmmakers started experimenting with changing individual frames of film and replacing them with images. For example, by shooting something in half of a frame and then adding in a plate to other half of the shot, you could get a spectacular effect—such as Charlie Chaplin skating on the edge of an abyss. The "abyss" was actually something added in later. Or you could shoot in two different parts of the frame and by just covering the lens in one section, rewinding, and then shooting in the part that was covered. The sky was the limit in the early days of cinema, and many filmmakers would experiment with these techniques in different ways. By 1933, special effects had advanced in bits and pieces. *King Kong* would push it all forward.

Willis O'Brien was the legendary special effects director in charge of the film and harnessing all of these techniques. He used a combination of miniatures—with specially-constructed steel armatures inside the puppets—and then used stop-motion animation to move the puppets to get life-like action. For stop-motion animation at 24 frames per second, O'Brien moved the puppet 24 times for one second of film. He would snap a picture, move the puppet, snap another frame, move the puppet, and so on. What made *King Kong* so remarkable was how realistic the action was, and in addition, how that animation worked in conjunction with the live action aspects of the film. Going back and forth between live action and animation was no small feat, and O'Brien mastered the technique.

Chapter 9: *King Kong*   151

Another technique pioneered by O'Brien was rear projection. He would build miniature screens into the animation sets, and then rear-project an element of live action into the scene, matching it up with the animation. The rear projection had to be done frame-by-frame, so when it played at full speed, everything was synchronized. O'Brien also used compositing to create certain sequences, such as exposing a sliver of the frame with one part of the action and then exposing the remainder of the frame with the rest of the action. In other scenes, he combined everything and exposed the film multiple times—one pass might have the live action, another pass might add the animation, and another pass might add live or static backgrounds. He also combined animated characters with real characters in the same frame, interacting with one another. (One famous side note is how Kong's fur would look like it was moving in the wind, when in actuality it was made by the hands of the animators moving the puppets between frames. O'Brien would lament this is as a mistake, but most believe it adds life to the puppet.) What O'Brien was doing was green screen/CGI work before there was a green screen or CGI, and doing it all in-camera in painstaking detail. For the fight between King Kong and the T-Rex alone, the sequence took seven weeks. If one frame was off or a simple mistake was made, they had to return to the beginning and do it all over again. The entire package was a remarkable feat, and it would inspire the field of special effects and animation for decades.

One sequence highlights this brilliance: after Driscoll is deposited by Kong inside his lair, we get a layered sequence that includes live action in Driscoll trying to escape, rear-projection in miniature behind him, a composite that puts animated Kong in the shot with him, and painted backgrounds to go along with the live-action and rear-projected backgrounds. There are several different techniques being used in this one sequence, and we see this throughout the film over and over again. It is groundbreaking brilliance.

While the film doesn't have aliens, spaceships, nor does it leave the Earth, it is still quite squarely in the realm of science fiction. Dinosaurs roaming the Earth long after they supposedly went extinct. Giant apes living next to humans. The seeds of *Planet of the Apes* and *Jurassic Park* are planted here, not to mention monster movies such as *Godzilla*. And just the way the film itself was made would inspire countless more science fiction films, moving beyond the realm of what was probable toward the realm of what was possible (and beyond). To the question of "What does it mean to be human?" the film answers this in a few fascinating ways. First it tells us that, unfortunately, exploitation is part of the human experience. It also warns that unfettered greed can be dangerous when unleashed (also see *Alien* or *Jurassic Park*). And finally, going back to my earlier point of how we start to feel about Kong toward the end of the film, and especially when he is being shot down: we feel sympathy. To be human is to identify not only with other humans, but with other living creatures of all kinds. We even feel love for those beings. To be human is to love.

—Vincent Piturro

Gigantic Animals

Jeffrey T. Stephenson

When I experience *King Kong*, I see it as groundbreaking or trendsetting in a number of ways: in its dinosaurs, in the trends the moviemakers were setting, and in bucking trends.

Gigantic animals far larger than the original fossil bones or known specimens is trend setting. Super-sized dinosaurs and other animals would become the norm for sci-fi following *King Kong* all the way through the latest offering of the *Jurassic Park* franchise. Well before motion pictures, and for many decades before *King Kong*, paleontologists were chasing the "biggest, longest, tallest" in a quest that was more for the superlative than for science. While gigantic long-necked Sauropods have indeed been discovered (and the largest ever let slip away!), there are no 80-foot long Stegosaurs.

Paleontologists have gone back and forth on the true nature of dinosaurs over the past 180 years: were they sluggish cold-blooded reptilian brutes, little more than overgrown lizards, or were they creatures that are more dynamic? *King Kong*'s dinosaurs do buck some of the trends popular in its day, especially those of the late 19th and early 20th century scientists who thought of these animals as brutish lizards doomed to extinction. The Kong dinosaurs still drag their tails, but otherwise have a much more dynamic set of behaviors really only popularized in the late 1960s by modern paleontologists.

King Kong also evokes some great earlier stories such as Jules Verne's *Journey to the Center of the Earth* (1864), Arthur Conan Doyle's *The Lost World* (1912), and Edgar Rice Burroughs' *The Land That Time Forgot* (1918). These sci-fi visits are similar to *King Kong*: each involve treks to the unknown, the uncharted, for discovery of the new and fantastic.

Before the Renaissance, maps of the known earth often had areas left blank except for the legend "here there be monsters" or perhaps a drawing or two of the creatures waiting to destroy those who dared sail—or sail into—those dangerous waters.

Mythical creatures of ancient lore are awesome, monstrous, hazardous, magical, and mysterious. Some ancient tales may be based on earlier humans looking at fossil remains and trying to bring these animals back to life in their imaginations. In my experiences with teaching natural history to age levels ranging from pre-school to senior life-long learners, three- and four year olds have a much easier time interacting with "live" dinosaurs. I often wondered how the Tyrannosaurus or Stegosaurus whose bones I helped excavate here in Colorado walked, ran, bellowed, ate, smelled—even as I was digging and plastering their bones. Despite my scientific training, I would daydream a bit on how Pteranodons would fly or Plesiosaurs would swim as I prepared their skeletons for study. Even as I helped dissect cadavers of Western Lowland Gorillas for scientific research, I would imagine what this animal might have been like, charging, feeding, bellowing, or fighting. I have worked with many paleontologists who have shared these reveries with me.

The opening of new places through exploration and colonization gradually narrowed the blank spaces of the old maps, but even after the turn of the 20th century there were places on the planet where new and unexpected creatures existed waiting to be found. "Living fossils" would sometimes turn up: the most celebrated of these was the Coelacanth, a fish thought to have gone extinct soon after the time of its Cretaceous-age fossils, only to turn up "in the flesh" in nets off Madagascar.

So why not have isolated places in remote areas of the world where remnant populations of unexpected creatures await discovery? As Carl Denim says to the Captain, "You won't find that island on any chart."

The Science

Science is exploration, science is discovery, science is figuring out how the world and its animals work and how they came to be.

There is something pure and honest about the original *King Kong*. The story relies on no secret formulae cooked up by a chemist gone mad. There are no covert or overt government conspiracies; no plans for genetic splicing; no plots by aliens, benign or otherwise.

King Kong is nature, raw nature, prehistoric nature, huge, mysterious, and menacing. It is without supernatural interference, as no ghosts or spirits or wild magic is involved (except in the speculation of one crewmember). *King Kong* is just natural. It is a great example of biological sci-fi.

One only needs to invoke natural evolutionary processes for the basics of Skull Island and its creatures, although I need to engage in "suspension of disbelief" when it comes to extinct animals from many geologic periods all coexisting on the same small island, and super-giant gorillas and dinosaurs far larger than anything we know. However, there is an explanation for gigantism in the hypotheses on Island Biogeography.

Islands are often refugia, places where species can survive sweeping changes of extinction on the mainland. Islands are where the very mysterious and seemingly prehistoric egg-laying mammals, the duck-billed platypus and the spiny anteater, survived. Islands had gigantic flightless birds such as the Elephant Bird of Madagascar, the Moas of New Zealand, and the Dodo of Mauritius. Sadly, humans drove many of these giants to extinction. Smaller animals isolated on islands grew into giant forms in prehistoric times, including shrews and rodents growing into the size of golden retrievers. Larger animals would sometimes get smaller when isolated on islands, including the seemingly non-sequitur pygmy mammoths from several different islands in the Pacific, and a truly tiny hippopotamus (much smaller than the Pygmy Hippo) from the Mediterranean.

The science of biology has been engaged in discovering and describing the biodiversity, evolution, behavior, and ecology of living and prehistoric creatures for about the last 400 years. Expeditions to wild and unknown places around the world uncovered strange and exotic creatures unknown to the general public. Excavations uncovered astounding, gigantic creatures with long sharp teeth and even longer deadly spikes from the fossil record.

Biology involves a large and varied range of studies from cryptozoology, genetic mutation, behavior, evolution, experimental manipulation, molecular and cellular systems, to discovery of new life forms. The era of biological discovery is not at all over, with as much as 80–90 percent of life forms not yet described by taxonomists. However, the golden age of this discovery, from the 17th–20th centuries, was one of new creatures coming from new countries or regions (or at least "new" to the western scientists who were describing them). These creatures included fantastic beasts such as platypuses, kangaroos, birds-of-paradise, and gorillas.

Dinosaurs and other prehistoric beasts were also being discovered during this time, and often described by the same scientists who were naming the modern animals. Charles Darwin was as much at home describing modern birds, barnacles, and beetles as he was with paleo mammals from Patagonia. Georges Cuvier, the famous biologist from France, described the Mastodon (named from the resemblance of its teeth to rows of female breasts), and thought it was extinct. On the other hand, Thomas Jefferson also studied mastodons and thought that these animals might still exist in the wilderness of interior North America. Jefferson even asked Meriwether Lewis and George Clark to look out for these beasts on their Journey of Discovery through the Louisiana Purchase in 1804–1806.

King Kong owes its scientific inspiration to the work of biologists and paleontologists and to their discoveries of many fantastic creatures.

Methods and techniques of discovering new living animals have changed or been augmented since the early explorations of the 18th century, with advances in a multitude of areas. New technologies allowed science to explore deeper, higher, in more remote settings, in the microscopic world, and on molecular scales. Recording discoveries started with quill and parchment but moved through many technologies to include photography, motion pictures, and the digital recordings of today.

The 1930s saw a rise in the adventurer-photographer who would bring their movie cameras along with their rifles and notebooks. This was true for many Natural History museums, becoming a mainstay much as dioramas or tours for school children. The Denver Museum of Nature & Science (back when it was called the Colorado Museum of Natural History) had its own naturalist moviemaker of the 1930s, Alfred M. Bailey, who amassed an archive of motion pictures of his explorations from Alaska to Campbell Island off Antarctica, and throughout much of the Pacific. Bailey would often present his films in the old Phipps Theater (since converted into an IMAX theater) to audiences from all over Colorado.

My job at the Museum is to care for, document, and make accessible the collections of the Zoology Department. We have over a million specimens of mammals, birds, insects, arachnids, marine invertebrates, and parasites. We have gorillas, as well as thousands of other species. When I am working with these specimens, none of them seem monstrous to me. So, what makes a monster?

Monsters

Definition of "monster" from the *Oxford English Dictionary*:
"Originally: a mythical creature which is part animal and part human, or combines elements of two or more animal forms, and is frequently of great size and ferocious appearance. Later, more generally: any imaginary creature that is large, ugly, and frightening."
Merriam-Webster:
"An animal or plant of abnormal form of structure."
Cambridge Dictionary:
"Any imaginary frightening creature, esp. one that is large and strange."
Definition of "monster" by "Judge John Hodgman" in *NYT Times Magazine* September 10, 2017, (p. 22) column "Bonus Advice," when asked about the difference between ghosts and monsters:

"'Monster' has connoted enormity and a certain savage verve since the 14th Century, whereas 'ghost' connotes a kind of sad floating around in corners." As no less an authority than Loren Coleman, director of the International Cryptozoology Museum in Portland, Me., put it: "Ghosts are not monsters. 'Monsters' are alive and may be biological." This posits the possibility that monsters can become ghosts and that the woods of the Pacific Northwest are haunted by spectral Bigfoots. But that awesome image is as close as your friend is going to get to erasing the meaning of words and paranormal taxonomy in my court.

The "monsters" cited in the question to Hodgman are the Loch Ness Monster and Chupacabra. Are monsters, by definition, enormous or gigantic? King Kong, Godzilla, the over-sized blown-up dinosaurs and other prehistoric reptiles in *King Kong* and Spielberg and much of the rest of sci-fi are gigantic, one might say monstrous. And it certainly doesn't hurt to be big if you're a monster. Even if that means that you are living well beyond all known exemplars of biological possibility.

However, if Chupacabra is also a monster, then size does not matter as much, for Chupacabra is drawn as being the same size as feral or wild dogs (almost certainly the inspiration for this very same monster), only extremely fierce—something with "savage verve." The specimens, on examination, turn out to be feral or wild dog carcasses that have lost much of their hair and jaws gaping wide showing their teeth due to the contraction of ligaments in the neck and jaws.

When one peers into a microscope at a tardigrade, the "water bears" only visible under very high magnification, they may appear monstrous, otherworldly beasts both ugly and ferocious-looking at the same time. There's something about them that just says "cute": perhaps a function of their miniature size. Would a 15-millimeter (just over ½ inch) Kong be "cute"? The 15-meter tall Kong is a "monster," and even a 1.5-meter Gorilla is "monstrous" to many with an incomplete understanding of our largest primate cousin.

Our Curator of Arachnology, Dr. Paula Cushing, studies spiders and solifugids. Many people dislike spiders but perhaps don't consider them monsters (unless, of course, you were the size of a fly). Black Widows are really quite willing to be left alone, and would rather take flight than bite. Most spiders are harmless. The solifugids (Camel Spiders, Wind Scorpions, Sun Spiders) do look evil, fierce, mean (our Curator refers to them as the Spawn of Satan). The largest Camel Spider is about the size of a mouse. If you were the size of a mouse, you might think a Camel Spider was a monster. But gorillas?

Gorillas do get to be big; a silverback male tips the scales at over 400 pounds and standing fully upright on its hind legs may be six feet tall. They are the largest living primate, and a close cousin of humans. They are almost exclusively vegetarians, only occasionally indulging in a side dish of insects, and do not eat mammal meat. The first reports of gorillas from the early 18th century described them as monsters, vicious beasts dangerous to life and limb. The silverback is in charge of protecting the group, and will charge and bluster, and if necessary fight to protect his kin.

"Monster" may need to be gigantic for fairy tales or Hollywood, or even for most people, but is plus size a necessity? If you were one-fifth as big as a water bear, you might think "Monster!" just as you might think the same, at your current size, when viewing a Rhinoceros. Overall, let's go with size does matter.

The Animals of King Kong

A few filed notes on the creatures of Skull Island, in order of sighting:

1. Kong, King Kong: (Gorilla sp. Novum)

- Looks very much like a Western Lowland Gorilla, only very, very big; over seven times larger than any known gorillas.
- Is fond of accepting female offerings from local village; is especially fond of Ona Mato Potato (blond females).
- Large foot prints, but apparently plantigrade (does not walk on the lateral sides of its feet like other gorillas). May be due to habit of walking upright most of the time.
- Grunts much like other gorilla species, but roar sounds like a lion.
- Curious.
- Very fast for a creature its size (>15 meters).

2. *Stegosaurus*: (*Stegosaurus* new species?)

- Looks like one of several species of *Stegosaurus* found in Jurassic rocks of Colorado and western North America, only very, very, very large; four times largest *Stegosaurus* known (over 24 meters long).
- Living fossil; last known relatives found in rocks 150–140 million years old.
- Belligerent and aggressive.
- Double-row of parallel plates on back (unknown in any *Stegosaurs*).
- Eight spikes on end of tail (again unknown in most *Stegosaurs*).
- Drags tail on ground (again unknown).

3. *Apatosaurus* (or *Brontosaurus*??): new species.

- One of the broad-skulled *Diplodocid Sauropod* dinosaurs of the Jurassic Morrison Age (approximately 150–140 million years ago), but very large; nearly twice the size of most *Diplodocids*. Note: may

actually be the missing dinosaur *Amphicoelias fragilimus* thought to be missing during transit from Colorado to New Jersey.
- Living fossil.
- Belligerent and aggressive, known to bite and throw people.
- Drags tail (unknown in *Sauropods*).
- Roar sounds like leopard.
- Teeth extend caudally all the way to the rear of the mouth (unknown in any dinosaur).

4. Paleozoic Arthropods, missing or unseen: (new genus and species)

- Reports sketchy, only second-hand accounts with no sightings, apparently giant spider- and millipede-like creatures similar to giant arthropods from 280 million years ago. No confirmed sightings; need to return.
- Thought to be in deep chasm on Skull Island.

5. Two-legged reptile: (new genus and species)

- Unusual cross, possibly hybrid?, between Tuatara-like and Chamaeleon-like reptiles, only very, very, very big.
- Living fossil? None known from fossil record.
- Climbing habit.
- Two claws on each of two front legs, no hind legs noted.

6. *Tyrannosaur*: (*Tyrannosaurus* new species)

- Looks very much like known *Tyrannosaurs*, late Cretaceous (c. 70–66 million years ago), but 30 percent larger than largest known.
- Three claws on hands (unlike all known *Tyrannosaurs*).
- Drags tail (unlike).
- Living fossil.
- Belligerent and aggressive.
- Growl sounds like mountain lion.

7. Long-necked *Plesiosaur* (*Elasmosaur*): (new genus and species)

- Looks like several mid–Cretaceous swimming long-necked reptiles, about the usual size.
- Four flippers, long neck.
- Living fossil. (Dates range from c. 100 million years ago–70 million years ago.)
- Belligerent and aggressive.
- Lives in scalding-hot volcanic mud puddles (really unusual for an *Elasmosaur*).
- Prehensile tail and neck used for strangling prey (also really unusual).

8. *Pteranodon*: (*Pteranodon* new species)
- Looks like other mid–Cretaceous *Pterosaurs*, but very, very big: over four times expected size.
- Large crest on back of skull.
- Opportunistic feeder.
- Wings like other *Pterosaurs*, but much shorter than *Quetzalcoatlus*; even though body twice as big as *Quetz*. Hard to understand how it achieves lift, let alone flies.
- Capable of lifting several dozens of kilos—far more than any known *Pterosaur*, ever.
- Belligerent and aggressive.
- Living fossil. Most known relatives from about 80–70 million years ago.

Falling Off the Phallus of Civilization: On Max Steiner's Soundtrack to *King Kong*

Roger K. Green

During the 1920s, the grammar of film had been developed by early masters of montage such as Sergei Eisenstein. Silent film at the theater, however, was far from silent. Theaters were filled with extra visual and olfactory sensations, and pianists worked in shifts throughout the day providing musical accompaniment, no two playing the exact thing the same way. Take the memoirs of one pianist for example:

> There was nothing passive about the audiences, especially the kids. To heighten the dramatic effect of tender love scenes, or to provide live sound for Westerns or battle scenes, the older kids would fire off the then popular Kilgore repeating cap pistols. The younger kids, identifying with the hero as he is being stalked by or about to be stabbed or jumped by the villain and his hirelings, would utter hysterical warnings like "Look out. He's behind the door!" There were always kids reading aloud the florid, polysyllabic subtitles to their mothers or grandmothers. At critical points, the film would split. This set off an orgy of applause, howling, banging, floor-kicking, whistling, etc. The audiences seemed to enjoy these "breaks" more than the picture.[1]

Unless you go to a late-night screening of *The Rocky Horror Picture Show*, you're unlikely to get such lively audience participation at the cinema these days. Aldous Huxley's science fiction classic, *Brave New World* (1931), appropriately projected a possible future for cinema by moving beyond the recent introduction of "talkies," films with synchronized soundtracks. Huxley's novel introduced the "feelies," where all of the audience's senses would be stimulated for its entertainment, long before virtual reality headgear. Huxley's book also resonated with a common theme occupying the cultural milieu of the early 1930s: savageness and civilization.

Max Steiner's soundtrack to *King Kong* (1933) remains pivotal in the history of film because of its unique blend of diegetic and nondiegetic sound. Composed at the expense of the film's producer, Merian C. Cooper, Steiner employed an unprecedented eighty-piece orchestra for his original music. In conventional film terms, *extra diegetic* refers to the sounds the audience hears outside of the staging and characters on screen, but what makes Steiner's work especially compelling is its complex contribution to storytelling of the film itself. Try watching *King Kong* with only subtitles and no sound and you'll quickly see how much the story relies on Steiner's score. We often take for granted the complexity of music in films precisely because of the mastery of composers like Max Steiner but also because when we watch a film, we generally do not see the musical performers or their instruments. Their very presence is often extra diegetic.

One of the techniques Steiner helped to popularize is known as "mickey mousing." Although Mickey Mouse was only five years old at the time *King Kong* was released, animated cartoons offered a more controlled correspondence between dramatic action and timed sound that would pave the way for sonic techniques in the production of talkies. It is not surprising that Steiner, along with Carl Stalling, who composed the music for the first Mickey Mouse film, *Plane Crazy* (1928), is credited as an innovator of the "click track." A click-track is basically a separate soundtrack containing the beat of a metronome to keep time, allowing editors to more easily align the sonic events with the actions on screen. Films are essentially rhythmic because it is merely the movement of still images through time that gives the audience the impression that the characters and objects in the frames move. Editors often cut to music.

One of the most interesting aspects of *King Kong*, however, is the *absence* of music throughout the beginning scenes, following the initial overture. The completely diegetic sound lends a kind of realism to the scenes in New York. It is not until the boat approaches Skull Island that we begin to hear music. As Robynn J. Stilwell writes, "Music rolls in with the fog, a visual metaphor for the fantastical gap. It is not yet the elsewhere of Skull Island, but an amorphous border that extends around it, blurring its edges."[2] We see a blending between diegetic and nondiegetic sound, foreground and background of the shots, and the environmental conditions of the sea and air. As Stillwell hints with the concept of the "fantastical gap" that locates the "geography of the soundscape,"³ this signals a kind of liminal space that draws both the characters on screen and the audience into the atmosphere of Skull Island. This fantasy within the fantasy of the film is also temporalized by a return to nature's "past" where dinosaurs and giant apes fight for dominance.

On the island, we continue to get a blend of diegetic and nondiegetic sound. As Michael Slowik notes, "The music that accompanies the islanders' dance is not precisely diegetic, because it features instrumentation plainly not visible in the image or likely in that setting. Yet it is not exactly nondiegetic either, because it retains clear connections to the dancing and drumming."⁴

In a Eurocentric scheme, this temporally "other" place represents a pre-political "state of nature," which Thomas Hobbes described as *bellum omnium contra omnes* or a war of all against all. Largely developing from contact with the so-called "new world" or "discovery" of what would become America, Europeans narrated a self-serving story of "civilization" that underwrote their claims to political and geographical domination worldwide. *King Kong* develops out of this implicitly racist line of thinking.

One of Merian Cooper's reasons for spending a lot of money on the production of a composed soundtrack for *King Kong* was an anxiety over the stop-motion effects of the King Kong and Dinosaur puppets. They didn't seem real enough. Steiner's richly textured orchestra added an element of seriousness to what might otherwise be entirely unbelievable. It also introduced "modern" compositional concepts from Europe to the medium of popular film, shifting

In the aesthetic terms of Friedrich Nietzsche's *The Birth of Tragedy Out of the Spirit of Music*, which had analyzed opera, Steiner's blend of diegetic and nondiegetic sound hovers at the border of the Apollonian and the Dionysian. While both Greek gods were associated with poetry and music, Apollo became associated with rationality and measured, civic spaces of the polis. Dionysus, on the other hand, was associated with nature, violence, and unbridled orgiastic sexuality. In these aesthetic terms, the shift between diegetic and nondiegetic sound on Skull Island in particular signals a gradated liminality between the "primitive" and the "civilized," and it is within this metaphor that I ask the reader to ponder a question: To what extent can a film soundtrack be racist?

Critics of *King Kong* have often viewed its story in terms of the racial anxieties of white populations who feared the encroachment of darker skinned "others" on what they perceived as their territory supported by their civilization story. This most obviously plays out in the misogyny-tainted romance between Ann Darrow (Fay Wray) and John

Driscoll (Bruce Cabot). Their romance hovers between the creative exploitations of filmmaker, Carl Denham (Robert Armstrong), and the infatuation of Kong himself. Kong was played by the uncredited African American actor, Everett Brown, in an ape costume. On Skull Island, Kong demonstrates his dominance over the more "primitive" dinosaurs, and he is worshipped by the island "natives" whose movements and ritualized sacrifices are also obscurely mirrored between the diegetic mickey-mousing and nondiegetic instrumentation of "tribal" drumming and orientalist textures. Whether in the jungle or atop the Empire State Building, Ann Darrow's "beauty" assuages the otherwise bestial fury of Kong the "beast." Her gruff and dismissive "common man," the sailor becomes the appropriate sexual alternative and heroic partner. Darrow's (white) "beauty" and supposedly intrinsic elegance gives pause to Kong, offering a temporal delay in violence such that the forces of "civilization" may tame or—in the final moments of the film—dispose of his bestial threat to her and the city of New York.

    I am going to assume that the reader intuits what I have explored in more depth elsewhere[5] as the "European phantasy" of civilization at work at the level of the plot: modern humans are alienated from nature due to their rationality; men are more rational than women, so women are "closer" to nature; children are also closer to nature and therefore more like "primitive" cultures as they developmentally progress toward rational, civilized adulthood, etc. But if we think sonically, Where was Steiner getting all of his material for suggesting these complex metaphors of liminality? What exactly was his contribution to this collaborative endeavor? Well, it is partly from his own Viennese upbringing. Steiner had studied with Gustav Mahler before coming the United States, and he was well-equipped with twentieth-century innovations in European composed music that may have seemed esoteric to American film-going audiences. He had also studied the operatic work of Richard Wagner that had inspired Nietzsche in *The Birth of Tragedy*. In one interview Steiner says, "Richard Wagner would have been one of the greatest picture composers that ever lived because he was underscoring dialogue just like I do. They talk. They have these endless adlibs, if you know what I mean.... What the hell is that but underscoring? The same thing I was doing."[6] Nineteenth century opera was partly based on European imaginings of what Greek music and tragedy were like. As Nietzsche writes in *The Birth of Tragedy*, the figure of the satyr—a companion to Dionysus—blended the rationally human and animalistic qualities. Underscoring hovers at the liminal space between diegetic and nondiegetic, but it also carries a Greek-derived baggage about a line between the civilized and the barbarous. As Nietzsche wrote, "A particularly modern weakness inclines us to see the primal aesthetic phenomenon in too complicated a way. For the true poet the metaphor is not a rhetorical figure but a representative image that really hovers before him in place of a concept."[7] John Driscoll's working class roughness and blatant misogyny situate him between Kong and Carl Denham, but it is the soundtrack that allows the plot to move the audience into the necessary disbelief that simplifies the "primal aesthetic phenomenon." Driscoll is merely a safety net, a place-holding device that allows the audience to enter into a "primal" fascination with Kong and Ann Darrow.

Composer David Allen points to a remark by Royal Brown insisting that "Steiner's *King Kong* music … would no doubt have scandalized most concert-going audiences of the time with its open-interval harmonies and dissonant chords, its triton motifs, or such devices as the chromatic scale in parallel, minor second."[8] While it is true that twelve-tone music was relatively new to the United States following works such as Arnold Schoenberg's *Theory of Harmony* (1910), which had broken away from diatonic (major scale) traditions that had shaped the "western ear" for centuries, we ought not confuse "concert-going audiences" with people who attended talkies. Kong and Ann Darrow offer the audience a kind of cathexis, the official impermissive attributes of their relationship are permitted by the plot of the film itself, which in turn gives the audience permission to indulge in the rawness of sexual violence. The blurring of the permissible and impermissible was sonically signified by chromaticism in the European phantasy structure.

A paradox presents itself if we think of the most avant-garde chromaticism of European music composed within the "classical" continuum, music which would presumably signal the most "advanced" evolution of "civilized" taste. In the European phantasy structure, "civilization" had moved *away* from the ape, the jungle, etc. At the same time, composers had long used aspects of chromaticism to signified and oriental "other." If we consider his biography, what Steiner was drawing on for a sense of cultural ambiguity and liminality may have been a sense of displacement that he himself felt as a displaced foreigner and a Jew. His status as an immigrant also combined with the musical textures he encountered when he came to the United States.

Watching *King Kong*, it is important to notice the long sequences where the music itself conveys the action, diegetic underscoring meant to persuade the audience to emotionally believe in the plot that the editor has crafted through images. We get with Steiner's score a sense of the "exotic" that will be repeated in his scores, such as *Casablanca* (1942). What is it about the harmonic and melodic content that conveys to audiences a sense of "otherness"? Orientalism in European music had been employed at least since the late nineteenth-century, notably in impressionist composers such as Claude Debussy and Erik Satie. Inspired in part by the late Romanticism of works such as Gustav Flaubert's *Salammbô* (1862) and in part by the 1889 *Exposition Universelle* (World's Fair) in Paris, these composers were directly concerned with the sonic representations of "other nations" underwritten by a kind of European cosmopolitanism. Such "universalizing" tendencies, which had drawn on representations of the "exotic," what Edward Said described in his famous book, *Orientalism* (1979), may have had a different impact on Steiner, who belonged to an already marginalized community within Europe. As an expatriated Austrian Jew, Steiner was himself a refugee. Facing discrimination as a Jew and a "foreigner" in England during the First World War, Steiner had been able to secure passage to the United States. His "otherness" would have been intimately tied to his sense of self. When we think of his composing hands at work, we can only imagine the layers of irony at work as he sonically signaled exotic "otherness" for his audiences.

Educated through the Austrian musical culture at the height of European modernism, as well as within a family-structure where work in live theater crossed the line between "high" and "low" art, Steiner himself negotiated the tensions between the culturally "other" and the "civilized." Interestingly, though little noted among historians of his work, Steiner entered an American musical scene where pioneers of jazz such as Duke Ellington had already been laying down a trans-continental texture of "jungle music" that would code a kind of accentuated resistance to more dominant strains of "primitivist" art. For Ellington, this would be part of an emergent impulse among the Harlem Renaissance artists in what would later be identified as a pan-Africanist aesthetic that turned primitivist essentialism against itself. Nathan Platte correctly notes that, "Steiner even drew upon the so-called 'jungle music' of Duke Ellington that was served up for white audiences at Harlem's Cotton Club. On Skull Island, a swung riff set over a walking bass blearily sustained chords gestures toward Ellington's 'East St. Louis Toodle-oo' (1927) and 'Echoes of the Jungle' (1931)."[9]

We know well that, conceived as cultural "outsiders," both Jews and African Americans influenced the phenomenon Americans call "jazz." Should we then think of the echoes of Ellington's influences in Steiner's compositions as homages? As the recognition by one cultural "outsider" of another? As aesthetic appropriation? What kind of complexities arise when a member of an oppressed group adopts hegemonic aesthetic notions? Can we presume to know anything about the intentional process, if there is any, at work here? Steiner's success with audiences is undeniable. Are audiences implicated in the injustices underscoring their enjoyment—not the composer's individual *intent* but the metaphysical ordering that normalizes our collective monstrosity?

In music law, there is a famous case involving a song by former Beatle

George Harrison from his album *All Things Must Pass* (1970). In 1976, Harrison went to trial for a dispute over his hit, "My Sweet Lord," regarding its resemblance to a 1962 hit by the Chiffons and written by Ronnie Mack, "He's So Fine." The judge found Harrison guilty of "subconscious plagiarism."[10] Such decisions are made by reference to recognizable melodic and harmonic structures. It is clear to any listener that "My Sweet Lord" closely resembles "He's So Fine," but what happens when we are dealing more with more textural concerns? This takes us back into the realm of aesthetics.

Like much music, film as a medium is well-suited to the discourse of psychoanalysis because films are composite works made by many people. Yes, some of those people have more power over the finished product than others, but there are many contributors. This makes any discussion of intention rather complex. Nevertheless, such complexity does not mean we ought to minimize issues such as racism and sexism, which are very much part of the fabric of *King Kong*. Nor does the question of intent, whether or not those involved are conscious or unconscious of the fact, decide the case. In fact, it is a rather weak sense of justice that can simply label a film (or novel, etc.) as racist and write it off, censor it, or ignore it.

The role of criticism is at least partly to try to see the film for what it is. This is not a matter of attempting an "objective" or disinterested stance. Criticism has its own discursive motivations, which far exceed whether or not we like or dislike a film. What remains compelling with *King Kong* is the iconic place it has culturally, especially the image of Kong and Ann Darrow atop the Empire State Building. At the time, the building was the tallest in the world, so the symbolism was quite powerful. The music in the scene accompanies Kong's steady ascent to the top of the world. After the climactic chest pounding, as Kong has put Ann down and climbed up to the antenna, the music disappears. We only get the sound of airplanes and gunfire until the fatigued and wounded Kong begins to sway. Horn blasts accompany the final shots, and Kong falls to his death. The ominous tones then merge into consonance as John reaches Ann. On the street below, Carl Denham tells a police officer, "It wasn't the airplanes, it was beauty killed the beast."

The success of John and Ann's romance is an afterthought, again a placeholder giving the audience permission to relish in Kong's perplexed lust for Ann. Clearly, we are meant to read Kong's death as a kind of tragedy. There is an element of questioning going on at this point in the film. If modern literature and film are characterized by treating the "common" person as the tragic hero, *King Kong* in its racist fantasy asks its audiences "how low can we go"? In its hyperbolic manifestation of civilization versus savagery, however, John Driscoll's misogynist love for Ann Darrow is rather boring. That Denham the filmmaker gives us the closing comment unites his artistic fascination with "the beast," but his very existence as an artist depends on his status as a colonizer, as a failed tamer of "the beast," as the adventurous entrepreneur only too willing to create a spectacle for his marvelous possession. There is little nobility to Kong's savagery, but he is perplexed by "beauty" in a way that reveals Denham the artist's own impotence. Denham seems to lament the loss of feeling and passion, projecting his own narrative of modern alienation onto "the beast" who has in reality been murdered by his own narcissism.

Kong has fallen from the phallus of civilization. The soundtrack renders him believable in his trans-species structure of delay. He is moved from a "timeless" past into a modernity that can only have its interest piqued by the surprise that there somehow still existed another savage to erase. In the meantime, boring misogyny is normalized as integral to plots and "happy" relationships from John and Ann to Walter Burns (Cary Grant) and Rosalind Russell in *His Girl Friday* (1940) and beyond! Kong becomes the sacrifice for the "moderns" that mimics the ritual sacrifice of the islanders, just bigger and better, and the soundtrack has seduced the audience into the believability of the unbelievable by suturing the diegetic and nondiegetic. How passive we have become as audiences.

Notes

1. Judith Crist, "Note for Abe Lass' *Play Me a Movie*," Liner notes for *Play Me a Movie Composed and Played by Abraham Lass*, Asch Records AH 3856, 1971, LP: 1–2. https://folkways-media.si.edu/liner_notes/folkways/FW03856.pdf.
2. Robynn J. Stillwell, "The Fantastical Gap between Diegetic and Nondiegetic," *Beyond the Soundtrack: Representing Music in Cinema*, eds. Daniel Goldmark, Lawrence Kramer, and Richard Leppert (Berkeley: University of California Press, 2007): 189.
3. *Ibid.*, 187.
4. Michael Slowik, "Diegetic Withdrawal and Other Worlds: Film Music Strategies Before King Kong 1927–1933, *Cinema Journal*, Vol. 53, No. 1 (Fall 2013): 2. pp. 1–25 https://www.jstor.org/stable/43653633.
5. See Roger K. Green, *A Transatlantic Political Theology of Psychedelic Aesthetics: Enchanted Citizens* (New York: Palgrave, 2019).
6. Myrl A. Schreibman, "On Gone with the Wind, Selznick, and the Art of "Mickey Mousing": An Interview with Max Steiner," *Journal of Film and Video* 56, No. 1 (2004): 46.
7. Friedrich Nietzsche, *The Birth of Tragedy* (New York: Penguin, 1993): 42.
8. Royal S. Brown, *Overtones and Undertones: Reading Film Music* (Berkeley: University of California Press, 1994): 118, quoted in David Allen, "'King Kong' by Max Steiner (1933) and James Newton Howard (2005): A Comparison of Scores and Contexts," davidallencomposer.com, February 7, 2014, accessed December 26, 2019 https://davidallencomposer.com/blog/king-kong-max-steiner-james-newton-howard-comparison#cite-4.
9. Nathan Platte, *Making Music in Selznick's Hollywood* (Oxford: Oxford University Press, 2017): 67.
10. Jordan Runtagh, "Songs on Trial: 12 Landmark Music Copyright Cases," *Rollingstone.com* June 8, 2016, https://www.rollingstone.com/politics/politics-lists/songs-on-trial-12-landmark-music-copyright-cases-166396/george-harrison-vs-the-chiffons-1976-64089/.

Chapter 10: *The Martian*

Matt Damon as Mark Watney stranded on Mars in *The Martian* (20th Century-Fox, 2015).

Ridley Scott is one of the great directors in the history of cinema, both in the sheer number of films he has made as well as the breadth and success of them. As the director of *Alien* (1979) and *Blade Runner* (1982), his reputation as a master of the sci-fi genre was solidified a long time ago. But he has made many more films throughout his career, of all types and in all genres, and his entire body of work is quite memorable, expansive (22 feature films), and successful. Many of his films as director are household names: *Thelma and Louise* (1991), *Gladiator* (2000), *Hannibal* (2001), *Black Hawk Down* (2001), *American Gangster* (2007), *Robin Hood* (2010), *Prometheus* (2012), and *The Martian* (2015). He has also produced many others, including *Blade Runner 2049* (2017) and dozens of TV shows among his over 100 producing credits. Even though he is now ensconced as one of the great directors of cinema history, he started in television and came to film relatively late, at the age of 40.

Born to an army family in northeast England in 1937, he attended the Royal College of Art in London, graduating in 1963. He worked exclusively in TV for a long time, ultimately forming a production company in 1968 with his brother Tony (also a director). He was a fan of science fiction literature as a kid, and he has said he enjoyed all of the science fiction cinema of his era, including *The Day the Earth Stood Still* (1953) among others. When he saw *2001: A Space Odyssey* in 1968, however, he became convinced the he too could make great science fiction films someday. After his first feature film, *The Duellists* (1977) was well regarded at Cannes, he got the opportunity to do just that in 1977 when he was offered the chance to direct a small sci-fi film called *Alien*. He accepted.

Alien was already in pre-production in late 1977 when *Star Wars* became one of the biggest blockbusters of all time. Due to the runaway success of that film, Scott lobbied for a bigger production; the budget for *Alien* was subsequently substantially increased and it became an A-list production for the studio. The success of the film solidified science-fiction as a box-office draw, showing that science fiction films aimed at adult audiences could make money and become critical successes as well. Sci-fi was going through a golden age.

The film owes much of its look to Swiss artist H.R. Giger, who designed the title creature as well as much of the film's environment. Much like George Lucas in *THX 1138* a few years earlier in 1971, Scott wished to depict a "used-up" or "retro-fitted" future and not one that was clean and sanitized, as in *2001: A Space Odyssey*. The film's dark, brooding atmosphere and stylized look became a prototype for the postmodern science fiction film and inspired countless other films. It also spawned several remakes and became one of the more successful franchises in film history. One aspect of Scott's work in this film that would stand out is the *mise-en-scène*; Scott proved to be an adept hand at creating the atmosphere through these details, and it would become a staple of his work throughout his entire career. The cinematography would always work hand-in-hand with the other aspects, but it would be the *mise-en-scène* that would stand out. As with the other films in this collection, the film's cinematic aspects work together to create the atmosphere, illuminate its themes, and propel the narrative in wonderful ways.

The story is very simple: in the far-away future, an industrial mining ship on its way home to Earth stops at a passing planet to investigate a possible life form. What they find is a deserted ship and an alien creature that begins to terrorize the crew. The film would become famous for many reasons: the degraded *mise-en-scène* of the future, not the sleek and clean view of Kubrick's vision; the abject capitalism of the world, one in which the people are simply dispensable cogs; the class and race issues depicted on the ship, very different from the inclusive *Star Trek* view of the future; and a strong female lead who becomes a tough killer. It was Scott who changed the original concept of a male lead to the female lead. The film would go on to make well over $100 million and instantly made Scott an international directing star. His next film, however, would solidify his place in the canon of the great sci-fi directors.

After the runaway success of *Alien*, he moved directly into the production for *Blade Runner*, based on Philip K. Dick's *Do Androids Dream of Electric Sheep?* Scott was originally hesitant about directing the film because he did not want to be pigeonholed as a sci-fi director after *Alien*, and then he differed with the studio on many issues surrounding the film. He finally finished his first cut in 1981. That original cut of the film was never released because the studio hated it, and they re-cut the film with a voice-over and an alternate ending. It was a box-office bomb, but it gained a cult following on video/laserdisc(!) in the following years. A Director's Cut version was released in 1992 and the film quickly became a huge hit after that. (That Director's Cut was the original version of the film, sans voice-over and tacked-on ending.) Yet another version, the Final Cut, was released in 2007 and Scott finally got the chance to have complete control and present his vision. It is now considered one of the most innovative, influential, and visionary science-fiction films of all time.

One reason for the success of the film is the vision of the dystopic future from the vantage point of the early '80s, and as in *Alien*, it is the *mise-en-scène* that stands out. The aesthetic depicts a used-up, retro-fitted future much like elements from *THX 1138* and *Alien*. But where those two films had a limited scope of vision, *Blade Runner* extrapolates the landscapes onto all of futuristic humanity. The vision of future Los Angeles—dark, dirty, dingy, and littered with *stuff* (what Dick called "kipple" in the book)—would become iconic images in sci-fi. It also pays homage to *Metropolis* in its verticality and how that height metaphorically relates directly to class and socioeconomics. The film is also (mostly) faithful to the beloved novel while adding its own take on the atmosphere of the future. Philip K. Dick, after seeing an early cut of the film, said: "They did sight-stimulation on my brain." Unfortunately, Dick died before the film was released. Scott was now both a commercial and artistic success, and he would have his choice of films for the rest of his career.

He chose not to return to sci-fi until 2012 with *Prometheus*—a spinoff of the space jockey from *Alien*. The film was a moderate financial success, and it divided both critics and audience. He would then return once again to sci-fi with *The Martian* in 2015 and have better results on all fronts. Based on the book by Andy Weir, it tells the story of astronaut Mark Watney (played by Matt Damon in a wonderful, virtuoso performance), who is presumed dead on Mars and is subsequently left behind by the rest of the spaceship Hermes crew (captained by Jessica Chastain) when they have to leave in the middle of a dangerous storm. Marooned by himself, he improvises a way to grow food and keep himself alive (and communicate with Earth) as the Earth crew devises a way to save him. With the next drop on Mars not scheduled for four years, his prospects are dire (especially after an accident wipes out his ability to grow any more food) until the Earth crew finds a way to send the Hermes back to pick him up. The gambit works and he is returned to Earth. The film was a critical and box-office success, and it was nominated for seven Academy Awards, including Best Picture and Best Actor. Along with the *Arrival* nomination for Best Picture the following year, and the multiple nominations for *Interstellar* the previous year, "adult" science fiction was experiencing a new golden age and was finally getting recognized not only at the box office, but at the Academy Awards as well. Ridley Scott's wish to bring adult sci-fi to the screen with *Alien* in 1979 was realized once again, and this time, others were on board as well.

The film itself is a master class in filmmaking (and acting); the rare film that appeals to a wide audience in a number of areas: story, acting, presentation, technical and special effects, as well as our four constituent elements of film: *mise-en-scène*, cinematography, editing, and sound. As with other Scott films, and particularly his science fiction films, the *mise-en-scène* is particularly expressive and it helps to tell the story in many ways. The opening shot is one great example: we see Mars, from low orbit, with the sun in the background. We cut to the surface and move along the rocky, mountainous terrain as if we were landing on the surface ourselves. The camera finally settles on the base camp and finds the astronauts busy working. The opening is pure Scott: economical, descriptive, yet also very expressive. The shot of Mars orients the viewer, but it also illuminates the solitary nature of the planet itself—lacking life of any kind. The terrain reminds us that Mars is a rocky, dusty planet and reinforces the idea that it is desolate and certainly dangerous. The shots of the astronauts, however, add life and *business* to the proceedings. Everyone is busy working and engaged in their specific tasks, but not without pleasure—their easy and even jocular joking banter show that they are a tight-knit group who care for one another on both professional and personal levels. Most of all, the astronauts add life to the planet and therefore give it a sense of community and togetherness. They bring Mars alive.

The cinematography, while striking in the opening shots, doesn't get in the way or announce itself in the way that Kubrick does, or even how Nolan (*Interstellar*) or Villeneuve (*Arrival*) use the camera as an explicitly expressive tool. Yet that fact also has its own expressiveness: the business-like quality of the shots—such as the opening sequence that opens the film in a very classic Hollywood cinema-like fashion—speaks to the business-like quality of the action; even though we don't get a specific time stamp for the film, throughout the course of the action we get the sense that it is very near-future, yet the Mars mission seems almost routine. And as the film progresses, another reason for the business-like approach of the cinematography becomes very apparent: the performance of Matt Damon. In the hands of another, less-experienced director, the temptation to become more stylized might have been too much to take, especially considering we are talking about a film *set on Mars*. There seems to be abundant opportunity to stylize the shots in many different ways. But Scott adroitly lets the acting and story do the work, and the camera allows it.

There are other ways, however, that Scott enlivens the action in a way that focuses on Damon but doesn't require him to do everything. Once he finds himself alone on the planet, Watney records a video diary of his activities and plans for posterity/science, and also in the hope that he is able to be rescued. Scott is able to use these sequences to give us different and unique views of Damon: some from the camera on which he is recording (so that it looks like the character is speaking directly to us), some from other cameras around him, and some from other surveillance cameras around the base. The net effect is that we get closer to the character—since we are mostly in close-up—but we also get a sense for how alone he is, all the time. Another of those surveillance-camera shots comes as Watney drives the rover around the surface. He records himself in the vehicle as he drives around, and we get the same type of access as when he is on the base. But these shots in the rover serve another purpose: his isolation and claustrophobia become piercingly evident. The small rover and his lack of mobility hammers home his situation, and despite his wise-cracking and joking (and dancing!), we understand his loneliness. At one point, driving around at night and trying to keep the heat off to save power, he is in the dark, his breath visible in the stark cold, and we realize his lot is one of not only loneliness, isolation, and claustrophobia, but stark danger as well. One wrong move, one inch, one rip in the wrong place, and he will die quickly. The cinematography doesn't dwell on this very much, but when we do get bits of it, the point is driven home in the starkest of terms.

The editing also works well with the *mise-en-scène* to make very expressive points. The overall scheme of the film is continuity editing—so that we are never confused about action, time, or place. In fact, we get screen titles every now and then to remind us of the day (listed by "Sol," the astronomical term for a Mars day). We also get screen titles for all characters on Earth, giving us their name and position. When we move locations, from NASA to the Jet Propulsion Laboratory (JPL) in Pasadena, for example, the screen titles tell us where we are, and the shots of planes give us the transition from place to place. All of this is very business-like and classic Hollywood cinema-esque. The viewer never gets confused.

The editing also becomes very expressive at times. As Watney makes contact with Earth and everyone realizes he's alive, the film moves between Earth and Mars more often. We get full sequences where the action cuts directly back and forth, but then we get sequences of cross-cutting (parallel action) that become very expressive in their overall feel. When Watney is first found to be alive, the action picks up on Earth as we see them following his movements. But later, when Watney has established contact and is in constant communication, the action on Earth becomes positively frenetic. When we cut to Watney, he is (obviously) alone and working in peace and solitude, making his own decisions. Cut back to Earth and NASA is always confounded with decisions and everyone disagrees with how to move forward: do they tell the crew of the Hermes Watney is alive? Disagreement. Do they attempt a risky mission that could put the Hermes crew in danger or rather, keep them safe and try a slightly less risky send of supplies so Watney can survive until the next mission in four years? Disagreement. Do they cut the safety prep time on the first mission to send supplies? Disagreement. Back on Mars, however, Watney moves along at his own pace, makes his own decisions, and works quickly and efficiently. At one moment, after NASA tells him to ration his food to barely a few mouthfuls, he looks into the camera and declares that he is dipping his potato (with which he is becoming disgusted with "the power of a thousand fiery suns") in crushed Vicodin, because he can and "nobody can stop me." The point here is clear: the bureaucracy bogs down and loses sight of the individual. Economic concerns trump human concerns. It may be a muted comment on capitalism and corporatism compared to *Alien* or *Blade Runner*, but it is there.

The sound also plays an important part in the film, and at times it is used for comedic purposes. Once Watney finds himself alone on the planet, he scours the stranded belongings for anything helpful/worth using. The only music he finds is Captain Lewis' (Chastain) cache of disco music on her laptop. The editing and music work in tandem in one simple cut when Director of Mars Missions Vincent Kapoor (Chiwetel Ojiafor) muses "I can't even imagine what he must be going through," and the film cuts to Watney exiting the shower in a towel listening to a blaring "Don't Leave Me This Way" by The Communards. Later, when he is making a journey in the rover, he extracts the Radioisotope Thermoelectric Generator (RTG), or as he calls it, the "big box of plutonium," to keep the rover heated. To pass the time, he listens to music along the way; his choice here is the "least disco song he could find," "Hot Stuff," by Donna Summer. The song choices are obviously on the nose, even a bit too cute perhaps, but effective nonetheless—especially when Watney starts to shimmy in the rover as the music blares. Who among us would not shimmy to Donna Summer on Mars?

The subjects of the film are deftly illuminated with the help of the cinematic aspects and some hearken back to Scott's older work in sci-fi. The subject of capitalism and the theme of the single-minded nature of the capitalist beast, for example, was a feature of *Alien*. That "beast" materialized in the form of the alien creature, whose only purpose was to survive and destroy anything in its wake. It even adapted to its surroundings and environment, taking on the physical properties of materials it encounters. It fed on those surroundings and found appropriate hosts. It was the physical embodiment of capitalism. The theme was also evident in the class divisions among the crew, as they all discussed their unequal division of "shares," while the captain reminded everyone that they would get what they "agreed to." The class division evident on the ship represented the supposed class divisions on Earth of the time period as well. While the film commented on the rising corporatism of the late '70s, it is still resonant in the bureaucracy of *The Martian*, as noted earlier, through the chain of command, but also in the budgetary issues faced in every step of NASA's plans. We see the venality of capitalism as a theme in *Blade Runner* as well, but we also see something else: the desire to be human. The replicants of *Blade Runner* simply want to become human and experience the life span that humans are afforded, but their maker, Eldon Tyrell, programmed them to have a limited life span (presumably to keep producing them and adding to his bottom line in perpetuity). While that film broaches the topic of wanting to be human, it also asks the question of "how do we retain our humanity in a world that is increasingly run by machines on which we are becoming increasingly dependent?" The Tyrell Corporation's motto for the replicants is "more human than human," and they certainly seem to be in the film. HAL, in *2001: A Space Odyssey*, as discussed in Chapter 3, has more emotion than the seemingly lifeless humans on the ship. The question becomes "How do we keep our humanity in a world that is increasingly based on machines?" And that consequently leads to the question we have been posing as central to science fiction cinema: "What does it mean to be human?" *The Martian* answers it simply: human connection. To be human is to connect with other humans; to want that connection; to need that connection; to even crave that connection. As Watney is alone on the planet, and he muses, "I am the only person to ever be alone on a planet," we understand his loneliness, and more, his lack of connection. Once he is able to communicate with NASA, it enlivens and invigorates him, and his plans and actions accelerate. When the crew finds out he is alive and they had left him behind, they do anything to save him. They put off their return to Earth for another year so they can go back and retrieve him. They too, not only needed to connect to Watney, they wanted him to hold on to his connections. To be human is to connect.

—Vincent Piturro

The Martian: Fact, Fiction or Fantasy? An Interview with Steven Lee

Ka Chun Yu

Ka Chun Yu: What is your background as a Mars scientist?
    Back when I was a senior in college (1976), I registered for an independent study class in planetary science. My professor (Dr. Joe Veverka) was involved in the Viking missions, intended to make the first landings on Mars by American spacecraft. Joe suggested I take a couple of weeks looking at all of the 7300-odd images of Mars returned by the Mariner 9 mission (the first spacecraft to orbit the Red Planet back in 1971–1972)—to see if anything caught my attention. What an introduction to another world! I saw a planet that was similar to Earth in many ways (mountains, canyons, polar ice caps, dust storms…) but also very alien (mountains nearly three times taller than Mt. Everest; a single canyon system that was 30,000 feet deep in places and could span the continental United States; polar ice caps covered with frozen carbon-dioxide [dry ice], and dust storms that can engulf the entire planet and persist for months). I was absolutely hooked, and from that point forward my entire career has focused on studying the interactions between the surface and atmosphere of Mars. Over the years, I've been lucky enough to have been involved in several spacecraft missions (Viking 1 & 2, Hubble Space Telescope, Mars Climate Orbiter, and Mars Reconnaissance Orbiter). In many ways, I'm more familiar with Mars than I am Earth!

What was your reaction when you first saw *The Martian*?
    I will admit up front that the first time I saw *The Martian*, I hated it! Being a Mars scientist, all I could see were the flaws where the movie had departed significantly from the reality of the Mars that I know and love. More about that later…

Chapter 10: *The Martian*   175

But now you no longer hate the movie?

The second time I watched *The Martian*, I tried hard to keep the mindset that this was *not* a documentary but a work of fiction. There are many interesting, entertaining, and well-done aspects to the movie! I'd like to touch on several of those. So my overall approach here is to discuss aspects of the movie that are fact, aspects which are fiction but plausible, and those parts of the plot that are pure fantasy.

So what is the first topic you want to talk about?

One of the basic underpinnings of the story is right at the beginning of the movie; a severe sandstorm is suddenly discovered to be bearing down on the *Ares III* landing site on Mars. Wind speeds are predicted to be above the limit where the Mars Ascent Vehicle (MAV—the spacecraft intended to carry the crew back into Martian orbit at mission's end) may be in danger of tipping over. The decision is made to immediately terminate the mission, evacuate the crew to the MAV, and blast off to meet the orbiting *Hermes* "mother ship" and begin the voyage back to Earth. During the hurried trek from the habitation module (the HAB) to the MAV, the crew encounters near zero visibility as they are engulfed by blowing dust, sand, pebbles, and rocks. The base's antenna array is ripped apart by the storm, and flying debris hits astronaut Mark Watney and blows him away into the storm. His damaged spacesuit no longer broadcasts any telemetry; after a frantic and fruitless search, his crewmates are forced to assume Watney is dead. Just in the nick of time, the MAV lifts off, successfully carrying the remaining crew into orbit. Several hours later, Watney awakes to "low oxygen alarms" and discovers he's been impaled by debris from the antenna array. He manages to hobble back to the HAB and….

Now—for the rest of the story! Let's talk first about the reality of this vicious "sandstorm" and how the wind can move surface materials on Mars. We have lots of observations (from Mars orbit and the Martian surface, as well as from telescopic observations from Earth) showing that dust storms are relatively common occurrences on Mars. They range in size from local, such as a dust devil that affects just a small area where the wind is circulating, to regional (covering hundreds to thousands of km), to massive global-scale events. I'd like to show you some of the real observations.

This is a view of the North Pole of Mars (Figure 1, right). The bright circular feature at top is the polar ice cap. There's also a swirling spiral-shaped cloud adjacent to the ice cap. That actually is very fine dust that's blowing off the polar cap and the surrounding surface. This is evidence for katabatic winds, where cold, dense air flows out from the polar cap onto the warmer surroundings. The winds lift fine dust present on the cap and surroundings, resulting in a turbulent cloud of dust moving away from the polar cap.

176   The Science of Sci-Fi Cinema

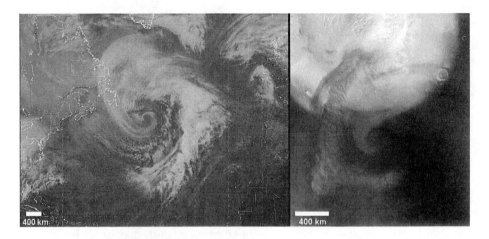

Figure 1 (left): METEOSAT-7 infrared view of a polar cyclone in the North Atlantic (ESA); (right): Mars Reconnaissance Orbiter view of north polar cyclone on Mars (MSSS/NASA/JPL).

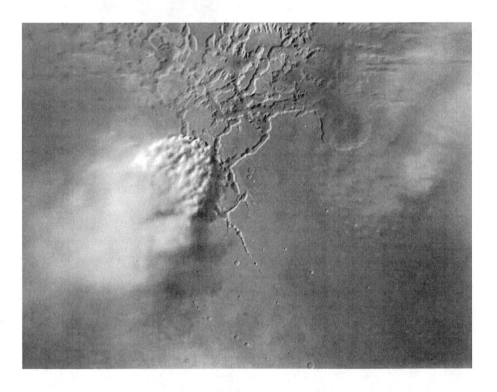

Figure 2: Regional dust storm near the Martian equator (MSSS/NASA/JPL).

This is not unlike what we see in the polar regions on Earth where polar cyclones are caused by the same sort of phenomenon (Figure 1, left). The difference is on Mars we've got dust swept into the cyclonic storm cloud, while on Earth condensed water vapor forms the clouds. But the basic process leading to the storm is very similar.

Elsewhere on the planet, heating of the surface by the sun can cause the overlying atmosphere to rapidly rise, much like thunderstorms on Earth (Figure 2). The winds caused by the upwelling air can lift dust from the surface, forming a dust storm tens-to-hundreds of km across. The prevailing winds can then carry the dust cloud across the surface, either lifting more dust along the way or slowly dispersing and allowing the suspended dust to settle back onto the surface.

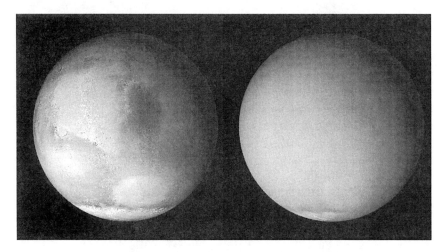

Figure 3: Global view of Mars in June 2001 (left) and July 2001 (right) (B. Cantor/MSSS/NASA/JPL).

This is an example of the extreme upper end of the size range for dust storms (Figure 3). It's called a Planetary Encircling Dust Event (PEDE), a.k.a. a global dust storm. On the left is a typical view of Mars; note a dust haze inside a huge crater (about 2300 km across) in the southern hemisphere. On the right, roughly a month later, the dust storm-grew large enough to escape the crater and get mixed up with the global wind field. At this point, the dust can be distributed throughout the atmosphere, forming a global dust haze (a PEDE).

So, Mars definitely hosts dust storms of a wide variety of scales. Based on observations by spacecraft which have been orbiting Mars continuously since Mars Global Surveyor arrived in late-1997, we see that on any given day there are dust storms happening somewhere on Mars.

When you talk about dust on Mars, it's not like what people's experience of dust is like, right?

Martian dust is much finer than the stuff you find under your couch. The average dust grains observed in the Mars atmosphere are 3 microns (a millionth of a meter) in diameter—about a tenth the diameter of a hair on your head. This is more like particles found in smoke. It's very, very, *very* fine particles! Once Mars dust grains are raised from the surface by the winds, it may take months for them to settle out of the atmosphere. In fact, some level of dust haze seems to always be present, leading to the famous salmon-colored sky ion most image from our landers and rovers. So once you're able to start a PEDE, it can be many months before the planet returns to a pre-storm level of atmospheric haze.

In fact, the most recent PEDE in mid–2018 led to the demise of the Mars Exploration Rover Opportunity. This rover had been operating for almost 15 years on the surface of Mars, when a PEDE kicked up. This was a dust storm of historic proportions; for several months, something like 99.99 percent of the sunlight was blocked by the pervasive dust clouds. This was bad news for the solar-powered Opportunity—essentially the middle of night, all the time, as far as its solar panels were concerned. The rover was unable to charge its batteries, leaving no power to turn on its internal heaters or communicate with mission control. Opportunity more or less froze to death in the dark! So, long-duration dust storms are big deals for spacecraft operating on the surface of Mars!

But the key here is—this is dust we're talking about! This is *not* sand, pebbles, or rocks blowing around (as shown in *The Martian*). The Mars atmosphere is very thin, only about 0.6 percent the density of Earth's atmosphere. (Sea level pressure on Earth is 1000 millibars; the average surface pressure on Mars is about 6 millibars.) It would take a truly enormous wind to lift anything coarser than sand off the surface. The force exerted by the wind is related to both the density of the atmosphere and the square of the wind velocity. Assume that to cause the mayhem shown in the opening minutes of *The Martian* you'd need the force produced by tropical storm winds on Earth (say, 100 km/hr). Our calculation shows this force would require a wind speed on Mars of about 750 km/hr (about 470 mph)! That's just physically unreasonable—it doesn't happen on Mars. We've explored Mars with landers, rovers, and orbiters; the maximum wind speed ever measured is about 100 km/hr. We also know that coarse material such as pebbles and rocks are not being moved around by the winds, and none of our landed spacecraft have ever been tipped over by wind gusts. So, the premise of the opening sequences of the movie can squarely be put into the "fantasy bin"!

The Martian sand or dust isn't just in the opening scene. It appears as set dressing later in the film as well.

That gets us to another point: Other than the plot device of requiring that the crew leave quickly because of this dire sandstorm happening, another task for Mark Watney is to figure out how the restore his communications with Earth. He decides to take a road trip to find and retrieve the Pathfinder spacecraft (landed on Mars in 1997), which is several days drive from the *Ares III* base. He finds Pathfinder almost completely buried by sand. And so the plot is again playing into the misconception that the winds are able to move a lot of material around on the surface over the course of a few decades. As we've already shown, that just doesn't happen under current conditions on Mars! Let's look at the types of changes that we *do* see on the planet:

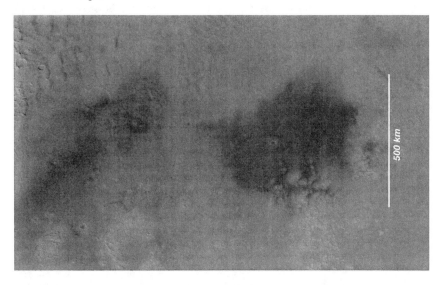

Figure 4a: Mars Reconnaissance Orbiter view of the Propontis feature, August 2009 (S. Lee/MSSS/NASA/JPL).

This is a view from the Mars Reconnaissance Orbiter (MRO) of the Propontis dark feature on the surface (Figure 4). It's about the size of the state of Colorado—about 500 km in each dimension. This feature has been visible from ground-based telescopes for the last 100 years; it's been a very stable feature. But in 2009, it disappeared over the course of a couple of months. The question is, what could have covered that up? When we go back and look at the daily images from MRO, there are dust storms persisting in the area. The storms are actually blowing over the dark feature for multiple days, and when the dust clouds finally clear or blow away the dark

180   The Science of Sci-Fi Cinema

Figure 4b: Mars Reconnaissance Orbiter view of the Propontis feature, December 2009 (S. Lee/MSSS/NASA/JPL).

Figure 5a: Opportunity rover, 2005 (NASA/JPL).

feature is essentially gone. That leads to the conclusion that some dust has settled down onto the surface. The dust is bright compared to the darker surface, and you've just covered it with a very thin amount of dust. That essentially erases the difference in the reflectivity of the dark feature and the brighter surrounding surface. What that also says is that even though you have these big dust storms, all they're doing is moving a thin veneer of dust around—they're not actually burying the surface deeply. We're probably talking about a fraction of a human hair thickness of material being deposited, not meters or even millimeters of dust.

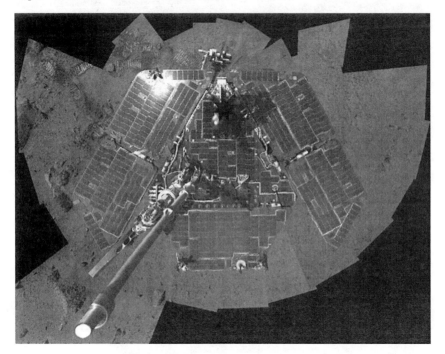

Figure 5b: Opportunity rover, 2007 (NASA/JPL).

Here is an example of how this process works. Back in 2005, the Opportunity rover took a self-portrait by looking down at the spacecraft with its mast-mounted camera (Figure 5a). The shiny dark solar panels that power the spacecraft are still squeaky clean at this point, even though more than a year has passed since landing. Two years later, you can see that dust has accumulated on the panels (Figure 5b), reducing the power generation by about 50 percent. However, you can still see the structure of the panels. Subsequently, winds presumably blew much of the dust away (Figure 5c). So it's an episodic process where you get some deposition, and then sometime later the wind regime has changed, perhaps in direction

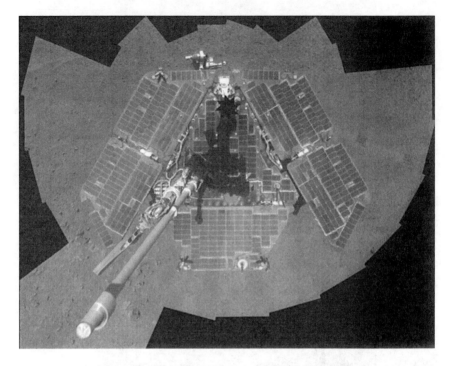

Figure 5c: Opportunity rover, 2008 (NASA/JPL).

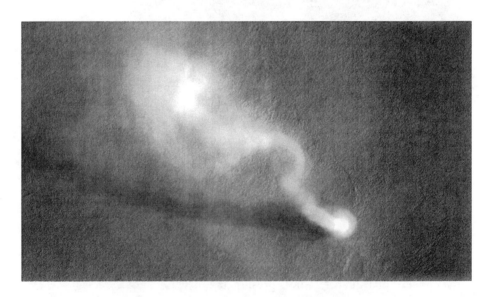

Figure 6: Dust devil on Mars (bright cloud). Measurement of the dark shadow indicates the dust devil is about 70 m wide and 20 km tall (MSSS/NASA/JPL).

or velocity, and it blows the dust away. So it's very hard to accumulate thick dust deposits—thick meaning millimeters or more. Substantial deposits probably take hundreds or thousands of years to accumulate.

Now we get into one of the processes that came as a surprise. Once we were able to operate landers or rovers on the surface for a long enough period, we actually saw that during some parts of the year dust devils sweep across the surface (Figure 6). These are very localized (tens to a few hundred meters wide), but if one of these dust devils goes right over the lander or the rover, it can clean the spacecraft off. The reason that Spirit and Opportunity lasted as long as they did (many years longer than their 90-day design lifetime), is that dust was deposited at a gradual rate, and then all of a sudden a dust devil passed over the spacecraft and removed most of the accumulated dust.

Remember the Pathfinder lander that Mark Watney dug up to scavenge the electronics from? This spacecraft landed back in 1997. MRO has imaged the landing site several times in the past ten years, and the spacecraft is still clearly visible sitting there on the surface. So obviously it's not been buried in the 20-ish years since the spacecraft landed. It's gotten brighter because of the dust sort of mantling it, but it's certainly not buried! Another example going back even further is the Viking missions which landed in 1976. MRO "sees" both Vikings still visible on the surface. Both landers are clearly "unburied" on the surface!

How realistic are the landscapes that they depict in the film with what Mars actually looks like on the surface?

One thing that's dealt with well in the movie are the various terrains that Mark Watney has to traverse. He needs to reach the already-landed *Ares-IV* MAV in Schiaparelli Crater, about 3200 km and a 60-day-drive away. So how realistic is it to have these very, very rough knobs and mesa and buttes on the surface of Mars (Figure 7a)? In fact, the movie actually tried to recreate the locales pretty well. Comparing a shot from the movie with a view of the Curiosity rover's exploration site in Gale Crater (Figure 7b), and allowing for artistic license, the resemblance is striking.

However, much of Mars is covered with fine sediments and craters, making a straight drive very difficult. But when he gets to areas like the walls of craters that are actually very steep, or areas with a lot of relief and very rugged with sharp features, Watney seems to slow down and become more cautious. Though speeding across Mars in the dark of night is downright crazy!

Now let's get to the part of the film that everyone remembers. What about Mark Watney becoming a potato farmer to feed himself?

Another one of the key story threads of this movie was Watney's creativity to find a solution to stretching his food supply before he starved to

Figure 7a: Background terrain featured in *The Martian* (20th Century-Fox).

Figure 7b: Curiosity rover view of nearby Martian landscape (MSS/NASA/JPL).

death. Being an expert botanist, he devised a plan to grow potatoes starting with those brought from Earth as "fresh" treats. He converted part of the HAB to a greenhouse, carried in buckets of Mars regolith from outside to act as soil, fertilize it with human waste, plant the eye sections from the fresh potatoes, then harvest, replant and eat, and so on. And he came up with a way to use the hydrogen and oxygen in the atmosphere and hydrazine rocket fuel from tanks in the discarded section of the MAV to make water so he could irrigate his crop.

In 2014, MRO made observations from orbit using an infrared spectrometer, which can give you an idea of the composition of the surface. And it flew over some areas that appear to be ephemeral water flows, possibly from subsurface ice or brine which may occasionally warm up enough to briefly wet the surface. This research team found very distinct spectral features with the signature of sodium, magnesium, or calcium perchlorate. The Mars Phoenix, which landed in northern artic, detected perchlorates when regolith samples were analyzed.

On Earth, we have a few examples of surfaces that are rich in perchlorate. One such area in Antarctica is called Don Juan Pond. Strangely enough, even though it's located in one of the coldest spots on our planet, this pond stays liquid throughout the year. When field teams examined this area, high concentrations of perchlorates were detected in the liquid water. The resulting brine has a depressed freezing point, so the pond doesn't freeze. As far as the movie goes, the unfortunate aspect about perchlorates is its toxicity for most biology, including plants and humans. On Earth, the adage is that any place that you find liquid water, you find at least algae or bacteria—you find life. It goes hand-in-hand. Don Juan Pond, however, supports little if any biology....

So, we've discovered that perchlorates do exist on the surface of Mars. They may exist at low levels over much of the planet. They're actually caused by a little bit of precipitation of water vapor in the atmosphere. That makes it much less plausible that Mark Watney could actually grow potatoes in the Martian regolith. If there are perchlorates at the *Ares III* site, it would certainly hinder the growth of Watney's potato plants. In defense of Andy Weir, author of *The Martian*, the existence of perchlorates was discovered well after the book was written and the movie was made. So you can't hold that against them. Mars threw them a curve ball!

So given everything you've said so far, what's your final assessment of the film?

I started out by saying I hated the movie because of the "gotchas" of the "sandstorm fantasy" and the existence of perchlorates in the regolith. The second time I went to see it, I came to appreciate this was fiction and not a documentary. Finally, I could just sit back and enjoy the creativity all the characters expended in overcoming the numerous hurdles Watney faced. With the exception of the "gotchas," much of the story is within the realm of possibility. The technology they're using is not necessarily something that we currently have "on the shelf" in terms of spacecraft or mission capabilities, but it's not "500 years in the future technology" either. It's the type of thing you can imagine existing in several decades, given sustained resources to design and construct (i.e., if Congress and the White House are 100 percent behind the program). Also, the many solutions Watney created, and the rigors of surviving in the hostile environment of Mars—those are plausible. The movie has lots of reality spread throughout. Some aspects are pure fantasy, but there absolutely had to be a plot device that would require the crew to depart Mars in a big hurry. Probably the most plausible event would be a solar flare, where you had to take off from the surface and get into the "mothership"—the *Hermes* had a radiation shelter. But that's also the type of thing that's not visually amazing, nor would it make for a really suspenseful "let's-get-out-of-here" moment. Having said all that, *The Martian* was a very enjoyable movie. Well done!

About the Contributors

Nicole L. **Garneau** is a taste scientist and public speaker. Her formal training in genetics and microbiology led her to the Denver Museum of Nature & Science where she served as the Curator of Human Health and the director of the Genetics of Taste Lab from 2009 to 2019. She is the founder of three companies related to her work studying taste science and is a keynote speaker, team-building consultant, and advocate for equality in science.

Roger K. **Green** is general editor of *The New Polis* and a senior lecturer in the English Department at Metropolitan State University of Denver. He is the author of *A Transatlantic Political Theology of Psychedelic Aesthetics: Enchanted Citizens*.

Charles **Hoge** is a scholar of monstrosity, dark folklore and Victorian culture. He teaches at Metropolitan State University of Denver, and has published on ludology in *Doctor Who* fanfiction (*Transformative Works and Cultures*), medieval influences in *I Am Legend* (*Reading Richard Matheson: A Critical Survey*), phantom dogs in *Under the Volcano* (*Malcolm Lowry's Poetics in Space*) and how the dodo's extinction haunts 18th century England's literary culture (*University of Toronto Quarterly*).

Steven **Lee** is a space scientist in the Experiences & Partnerships division at the Denver Museum of Nature & Science, and is a senior research scientist at the Space Science Institute in Boulder, Colorado. He received a Ph.D. in planetary geology from Cornell University. His research focuses on the interaction between the surface and atmosphere of Mars—primarily by mapping the patterns of wind-blown dust deposits across the planet utilizing spacecraft observations.

Andrew J. **Pantos** is an associate professor in the English department at Metropolitan State University Denver. He is a sociolinguist whose research and publications focus on language attitudes and cognitive processing, identity, discourse, sociophonetics, forensic linguistics, and gender and sexuality. He regularly teaches a number of courses, including phonetics, semantics, anthropological linguistics, language and society, morphology and syntax, history of English, and analyzing English.

Naomi **Pequette** is a space science experience developer at the Denver Museum of Nature & Science. She connects all ages to the latest space science research through innovative exhibits, engaging lectures, demonstrations, and live planetarium shows. She holds a BS in physics and astrophysics from the University of Denver and is a four-time graduate of Starfleet Academy, the temporal extension program.

About the Contributors

Vincent **Piturro** is a professor of film and media studies at Metropolitan State University of Denver and holds a Ph.D. in comparative literature. His areas of study include Westerns, science fiction, documentaries, Italian cinema, film history. He hosts an annual Science Fiction Film Series in conjunction with the Denver Museum of Nature and Science and the Denver Film Society. He also writes the film review column for a local newspaper, *The Front Porch*.

Joseph **Sertich** is the Curator of Dinosaurs at the Denver Museum of Nature & Science where his research focuses on dinosaur ecosystems. His field-based work is divided between the southern hemisphere and western North America. He is one of the primary researchers on the Madagascar Paleontology Project and works on other Cretaceous fossils in Africa, including Kenya and Egypt. In North America, he leads the Laramidia Project, working to uncover dinosaurs in the Cretaceous of Utah, New Mexico, and Colorado.

Jeffrey T. **Stephenson** has worked at the Denver Museum of Nature & Science for over 30 years as a collections manager, first for the Education Collections and then for the Zoology Research Collections. His work in zoology encompasses both invertebrate and vertebrate fields, mostly with insects, arachnids, mammals, birds, and parasites. He has also volunteered in paleontological and archaeological fieldwork and lab work, and he conducts experimental archaeology in ancient Egyptian studies.

Ka Chun **Yu** is a research astronomer and science communicator at the Denver Museum of Nature & Science. His research has included studying how digital planetariums can be used effectively for astronomy education, as well as observational star formation. He regularly gives live presentations about space sciences to the general public, has developed planetarium software to visualize the known universe, produced movies for planetariums, and helped develop space science museum exhibits.

Bibliography

Allen, David. "'King Kong' by Max Steiner (1933) and James Newton Howard (2005): A Comparison of Scores and Contexts." davidallencomposer.com. February 7, 2014, accessed December 26, 2019. https://davidallencomposer.com/blog/king-kong-max-steiner-james-newton-howard-comparison#cite-4.
Brown, Royal S. *Overtones and Undertones: Reading Film Music*. Berkeley: University of California Press, 1994.
Cohen, Jeffrey Jerome, ed. *Monster Theory: Reading Culture*. Minneapolis: Minnesota University Press, 1996.
Crichton, Michael. *Jurassic Park*. New York: Alfred A. Knopf, 1990. Print.
Crist, Judith. "Note for Abe Lass' *Play Me a Movie*." Liner notes for *Play Me a Movie*. Composed and Played by Abraham Lass. Asch Records AH 3856, 1971, LP: 1–2. https://folkways-media.si.edu/liner_notes/folkways/FW03856.pdf.
Gilmore, David D. *Monsters: Evil Beings, Mythical Beasts, and All Manner of Imaginary Terrors*. Philadelphia: Pennsylvania University Press, 2003. Print.
Hone, David. *The Tyrannosaur Chronicles: The Biology of the Tyrant Dinosaurs*. New York: Bloomsbury, 2016.
Johnston, Keith M. *Science Fiction Film: A Critical Introduction*. London: Bloomsbury, 2011. Print.
Kingsley, Elizabeth. "Jurassic Park." *And You Call Yourself a Scientist!* 17 August, 2009. Web.
Nietzsche, Friedrich. *The Birth of Tragedy*. New York: Penguin, 1993.
O'Brien, Christopher. "Modern Dinosaurs?" OurStrangePlanet.com. 28 May 2017. Web.
Platte, Nathan. *Making Music in Selznick's Hollywood*. Oxford: Oxford University Press, 2017.
Rudwick, Martin J.S. *Scenes from Deep Time: Early Pictorial Representations of the Prehistoric World*. Chicago: Chicago University Press, 1992. Print.
Runtagh, Jordan. "Songs on Trial: 12 Landmark Music Copyright Cases." Rollingstone.com. June 8, 2016. Accessed December 26, 2019. https://www.rollingstone.com/politics/politics-lists/songs-on-trial-12-landmark-music-copyright-cases-166396/george-harrison-vs-the-chiffons-1976-64089/.
Sanz, Jose Luis. *Starring T. Rex! Dinosaur Mythology and Popular Culture*. Bloomington: Indiana University Press, 2002.
Schreibman, Myrl A. "On Gone with the Wind, Selznick, and the Art of 'Mickey Mousing': An Interview with Max Steiner." *Journal of Film and Video* 56, No. 1 (2004): 41–50.
Slowik, Michael. "Diegetic Withdrawal and Other Worlds: Film Music Strategies Before *King Kong*, 1927–1933. *Cinema Journal* 53, No. 1 (2013): 1–25. https://www.jstor.org/stable/43653633.
Stern, Megan. "*Jurassic Park* and the Moveable Feast of Science." *Science as Culture* 13 No. 3 (2004): 347–372. Web.
Telotte, J.P. *Science Fiction Film*. Cambridge University Press, 2001. Print.

Index

Adams, Amy 11, 12
Alexa (Amazon) 60
Alien (film) 6, 12, 33, 47, 130, 167, 168, 169, 170, 172, 173
All Things Must Pass (film) 165
Alphaville (film) 5, 59
Amemiya, Yoji 42
American Gangster (film) 167
Arecibo Message 21, 119
Arecibo Observatory 19
Armstrong, Robert 146, 162
Arrival (film) 7, 11–30, 33, 84, 87, 116, 170, 171
Ashitey, Clare-Hope 74
Asimov, Isaac 4, 71
Asylum (film) 84
"Atmospheres" 57
Attenborough, Richard 125
Atwood, Margaret 83
Avatar (language) 27
Avatar (movie) 28

Back to the Future (I, II, and III) 28
Bailey, Alfred M. 155
Barbarella 5
Barry Lyndon (film) 50
The Battle of Algiers (film) 78
The Beatles 164
Benchley, Peter 123
Bentley, Wes 32
The Birth of Tragedy (book) 162, 166, 189
Black Hawk Down (film) 167
Black Swan (film) 78
Blade Runner 2049 (film) 12
"Blue Danube" 55
The Book of Eli (film) 86
Book of Great Sea-Dragons 143
Bradbury, Ray 4
Brave New World (book) 160
Bridges, Jeff 84
Brown, Royal 163
Burroughs, Edgar Rice 152
Burstyn, Ellen 32
Busey, Jake 32

Cabaret (film) 53
Cabot, Bruce 146

Caine, Michael 31
Casablanca (film) 163
Cast Away (film) 110
Centers for Disease Control (CDC) 108
CGI 36
Chaplin, Charles 3, 147, 148, 150
Chastain, Jessica 31
Chiang, Ted 12
The Chiffons 165
Children of Men (film) 8, 73–83, 86, 87, 128
Chrichton, Michael 125
Chupacabra 156
cinematography 6, 15, 33, 36, 37, 52, 53, 54, 55, 57, 74, 75, 76, 77, 98, 99, 102, 114, 115, 116, 125, 128, 168, 170, 171
Citizen Kane (film) 148
Clark, George 154
Clarke, Arthur C. 51, 52, 56, 59
A Clockwork Orange (film) 50
Close Encounters of The Third Kind (film) 5, 17, 54, 124, 130
Coleman, Loren 156
The Color Purple (film) 124, 130
The Communards 172
Conan Doyle, Arthur 152
Contact 8, 110–122
Cooper, Merian C. 9, 145, 146, 147, 149, 160, 161
The Cotton Club 164
Cretaceous Cloverly Formation 131
Cuarón, Alfonso 73
Cushing, Paula 156
Cuvier, Georges 154
Cyc 67, 68, 69, 70

Damon, Matt 31, 36, 42, 167, 170, 171
Dark City (film) 6
Darwin, Charles 154
Davies, Marion 148
Davis, Ernest 71
The Day the Earth Stood Still (film) 4, 17, 159, 168
Debussy, Claude 189
del Toro, Guillermo 73
Denver Film Society 1, 188

Index

Denver Museum of Nature and Science 2, 7, 8, 9, 188
Dern, Laura 125, 126, 140
De Sica, Vittorio 75
Dick, Philip K. 78, 169
Dionysus 161, 162
Do Androids Dream of Electric Sheep? (book) 169
Dr. Jekyll and Mr. Hyde (film) 3
Dr. Strangelove (film) 50
Don Juan Pond 185
"Don't Leave Me This Way" 172
Dothraki 27
Douglas, Kirk 49
The Duelists (film) 168
Dumas, Stéphanie 21
Dunkirk 37
Dutil, Evan 21

Eagle Nebula 117
editing 7, 12, 15, 16, 33, 37, 38, 52, 56, 57, 74, 76, 77, 84, 87, 88, 89, 98, 99 100, 102, 103, 115, 116, 125, 128, 135, 170, 171, 172
Edwards, Christopher 44
8 1/2 (film) 59
Eisenstein, Sergei 159
Ellington, Duke 164
E.T. (film) 5, 124, 130
Eyes Wide Shut (film) 50

Facebook 60
Faith, Madness and Spontaneous Human Combustion (book) 108
Fantasia (movie) 142
Fellini, Federico 5, 59, 98
Ferris, Pam 83
Figuier, Louis 143
Flaubert, Gustave 164
Forrest Gump 2, 110, 111, 116
Foster, Ben 84
Foster, Jodie 110, 111, 112
The Fountain (film) 78
Fox, Michael J. 110
Frankenstein (book) 139, 140
Frankenstein (film) 3, 67, 139–140
Frankenstein, Dr. 67, 139, 140
Frigs, David 95
Frontiers of Psychology 121
Full Metal Jacket (film) 50
Futureworld (film) 5

Garfield 64
Garneau, Nicole 8, 74, 80–83, 90–96, 100, 103–109, 187
Gasser, Michael 69
"Gayane Ballet Suite" 57
Giger, H.R. 168
Gilmore, David 138, 189
Gladiator (film) 167
Godard, Jean-Luc 5, 59

The Godfather (film) 124
Godzilla (movie) 4, 151, 156
Goldblum, Jeff 125, 126, 139
Goodfellow, Ian 65
Google 60, 61, 68
Grant, Cary 166
Gravity (film) 73
Green, Eva 84, 86, 91
Green, Roger 9, 159–166, 187
Gyasi, David 32

HAL 52, 53, 57, 58, 59–72, 173
Hallam Foe (film) 84
The Handmaid's Tale (book) 83
Hannibal (film) 167
Harrison, George 165
Hawkins, Thomas 143
Hayes Code 5, 50, 51, 146, 148, 149
Hearst, William Randolph 148
Hell or High Water (film) 84, 85
Hello Kitty 64
Her 37
"He's So Fine" 165
Hobbes, Thomas 161
Hodgman, Judge John 155, 156
Hoge, Charles 8, 137–144, 187
Hoover, J. Edgar 147
"Hot Stuff" 173
Hubble Space Telescope 174
Hurt, John 111
Huxley, Aldous 160

I, Robot 4
I Wanna Hold Your Hand (movie) 111
"The Iguanodon and The Megalosaur" 143
"The Immigrant" 147
"In the Court of the Crimson King" (song) 76
Iñárritu, Alejandro 73
Incendies (film) 12
Inception (film) 31
Indiana Jones and the Last Crusade (film) 124
Indiana Jones and the Temple of Doom (film) 124
Insomnia (film) 31
International Academy of Astronautics (IAA) 121
Interstellar (film) 7, 13, 31–48, 87, 116, 170, 171
Invasion of the Body Snatchers (film) 4, 59
Ishikawa, Yoji 42

Jakosky, Bruce 44
James, P.D. 74
Jaws (film) 5, 123, 124, 127, 128, 130
Jefferson, Thomas 154
Jeopardy (show) 60
Jet Propulsion Laboratory (JPL) 172
Joffé, Roland 111
Journey to the Center of the Earth (film) 152
Jurassic Park (film) 5, 6, 8, 9, 113, 123–144, 149, 151, 152

Index

Kawin, Bruce 2
Keaton, Buster 3
Keitel, Harvey 87
The Killing Fields (movie) 111
King Kong (film) 9, 55, 124, 142, 145–166, 189
Klingon (language) 27
Knight, Charles R. 131, 133
Knight, Wayne 126, 141
Kubrick, Stanley 8, 49–59, 73, 86, 110, 113, 169, 171

La La Land (film) 84
The Land That Time Forgot (film) 152
Lantieri, Michael 127
Leaping Laelaps 131
Lee, Steven 9, 174–186, 187
Lenat, Douglas 67, 69
Lennon, John 59
Let the Right One In (film) 37
Lewis, Meriwether 154
Lincoln, Abraham 67, 68
Lithgow, John 32
Loch Ness Monster 138, 139, 156
Logan's Run (film) 5
Lolita (film) 50
The Lost World (film) 152
The Louisiana Purchase 154
Lucas, George 124, 168
Lumière Brothers 150
"Lux Aeterna" 57

Mack, Ronnie 165
MacKenzie, David 32, 84, 85, 86, 89
Mad Max (movie) 111
Mann, Anthony 49
Marcus, Gary 71
Mariner 9 174
Mars Climate Orbiter 174
Mars Global Surveyor 177
Mars Reconnaissance Orbiter 176, 179, 180
Martin, John 143
The Martian (movie) 9, 13, 42, 43, 48, 167–186
The Martian Chronicles 4
The Matrix 6, 33, 71, 78, 86
Mazzello, Joseph 137
McConaughey, Matthew 31
McGregor, Ewan 84, 86, 94
McKay, Chris 44
Mean Streets (film) 87
Mecca, Dan 89
Méliés, George 3, 16, 54
Memento 31
METEOSTAT-7 176
METI 21
Metropolis (film) 3, 4, 129, 169
Metropolitan State University of Denver 7, 9, 187, 188
Michelangelo 76, 83

Microsoft 60
Miller, George 111
Mise-en-scène 6, 14, 15, 33, 35, 37, 52, 55, 57, 74, 75, 76, 77, 87, 98, 99, 102, 113, 115, 116, 125, 126, 127, 168, 169, 170, 171
Moby Dick (book) 98
Monsters (film) 86
Monsters: Evil Beings (book) 138, 189
Moon (film) 78
Moonlight (film) 84
Moore, Julianne 74
Morse, David 112
Mother! (fim) 78
"My Sweet Lord" 165
Mythical Beasts (book) 138, 189

National Science Foundation 118
Neill, Sam 125, 126, 137
Nietzsche, Friedrich 56, 57, 58, 161, 162, 166, 189
A Night to Remember (film) 53
Nolan, Christopher 31, 33, 34, 35, 36, 37, 38, 39, 171
Nolan, Jonathan 31, 36

O'Brien, Willis 142, 150
Ohkita, Takaya 42
Ojiafor, Chiwetel 172
Oliver! (movie) 59
Olsen, Mark 101
One Million Years B.C. (movie) 142
Opportunity (Mars Rover) 178, 180, 181, 182, 183
Ostrom, John 131, 135
Owen, Clive 73, 74, 80

Pan's Labryinth 73
Parasitology & Vector Biology (book) 104
Pequette, Naomi 8, 117–122, 187
Perfect Sense 8, 84–96
Pi (film) 78
Picasso, Pablo 104
Pine, Chris 84
Pink Floyd 76
Piturro, Vincent 1–9, 11–16, 31–39, 49–59, 73–80, 84–90, 97–103, 110–116, 123–131, 145–152, 167–174, 188
Plane Crazy (film) 160
Planet of the Apes 5, 17, 151
Planetary Encircling Dust Event (PEDE) 177, 178
The Polar Express (film) 110
Pontecorvo, Gillo 75, 78
The Prestige (film) 31
Primer (film) 97
The Prisoner of Azkaban (film) 73
Prisoners (film) 12
Project Phoenix 118, 119, 120
Proxima Centauri 117
Psycho (film) 50, 51

Index

Quenya (language) 27

Radioisotope Thermoelectric Generator (RTG) 172
Raiders of the Lost Ark (film) 124
Rain, Douglas 60
Reed, Carol 59
Renner, Jeremy 12
Renoir, Jean 75
"Requiem" 57
Requiem for a Dream (film) 78, 87
Richards, Ariana 137
The Road (film) 86
Robin Hood (film) 167
Robocop 6
The Rocky Horror Picture Show (film) 159
Roma (film) 73, 74, 80
Romancing the Stone (film) 111
Russell, Rosalind 166

Sagan, Carl 8, 18, 111, 112, 113, 114, 116
Salammbô 164
Sapir-Whorf Hypothesis 29
Satie, Erik 164
Saving Private Ryan (film) 163
Schoenberg, Arnold 78
"The Sea Dragons, as They Lived" 143
Selznick, David O. 145, 146
Sensory Perception: Mind & Matter (book) 95
"The Sentinel" 51
Sertich, Joe 8, 131-137, 187
SETI (Search for Extra Terrestrial Intelligence) 17, 111, 118, 119, 120, 121, 122
The Shape of Water (film) 73
The Shining (film) 50
Shlens, Jonathon 65
Sicario (film) 12
Skerrit, Tom 111, 118, 119, 120, 121, 122
Smith, Linda 69
Snowpiercer (film) 86
Solaris (film) 78
sound 7, 12, 15, 21, 33, 52, 56, 57, 74, 77, 90, 98, 99, 100, 116, 125, 126, 127, 128, 146, 149, 159, 160, 161, 170, 172
Soylent Green 5
Spartacus 49
Stalling, Carl 160
Star Trek 2, 5, 17, 46, 47, 54, 169
Star Trek IV: The Voyage Home (film) 17
Star Trek: The Motion Picture (film) 17
Star Wars 2, 5, 17, 124, 126, 128
Steiner, Max 160, 166, 189

The Stepford Wives 5
Stephenson, Jeffrey 9, 152-159, 188
The Sugarland Express (movie) 127
Summer, Donna 173
Szegedy, Christian 65

The Terminator 6
Thelma and Lousie (film) 167
Theory of Harmony (book) 163
The Thirteenth Floor 6
Thorne, Kip 31
Thus Spoke Zarathustra (book) 56, 58
Tom and Jerry (cartoon) 64
"A Trip to the Moon" 3, 16, 150
20,000 Leagues Under the Sea 3
2001: A Space Odyssey 6, 8, 49-72, 85, 113, 116, 168, 173

Unsworth, Geoffrey 53
Upstream Color 8, 87, 97-109, 128
Used Cars (movie) 111

The Valley of the Gwangi (movie) 142
Varnum, Michael 121
Verne, Jules 3, 152
Very Large Array (VLA) 114, 120
Veverka, Joe 174
Viking 1 & 2 (spacecraft) 174, 183
Villenueve, Denis 12, 13, 15, 102, 171

Walden (book) 101, 109
Wall-E (movie) 47, 86
Weaver, Sigourney 111, 118, 119, 120, 121, 122
Wells, H.G. 6
Westworld (HBO series) 31
Westworld (movie) 5
What Lies Beneath (movie) 110
Who Framed Roger Rabbit? (movie) 110, 111, 116
Williams, John 127
Withey, Brit 1
World Health Organization (WHO) 81, 107, 108
Wray, Fay 161

Y tu mama Tambien (film) 73
Young Adam (film) 84
Yu, Ka Chun 7, 8, 9, 16-22, 39-48, 59-72, 174-186, 188

Zardoz (film) 5
Zemeckis, Robert 8, 110-116